Science Starters!

Science Starters!

Over 1000 Ready-to-Use Attention-Grabbers that Make Science FUN for grades 6-12

Robert G. Hoehn

The Center for Applied Research in Education, Professional Publishing
West Nyack, New York 10995

Simon & Schuster, A Paramount Communications Company

© 1993 by
THE CENTER FOR APPLIED
RESEARCH IN EDUCATION
West Nyack, New York

All rights reserved. Permission is given for
individual teachers to reproduce the science
worksheets and illustrations for classroom
use. Reproduction of these materials for an
entire school system is strictly forbidden.

10 9 8 7 6 5 4 3 2 1

Library of Congress Cataloging-in-Publication Data

Hoehn, Robert G.,
    Science starters! : over 1000 ready-to-use attention-grabbers that make science fun for grades 6–12 / Robert G. Hoehn : illustrations by Robert J. Hubbard.
    p. cm.
    ISBN 0-87628-860-3
    1. Science—Experiments—Methodology   I. Title.
Q164.H638 1993
507.8—dc20   93-4728
                                                                       CIP

ISBN 0-87628-860-3

**The Center for Applied Research
in Education,** Professional Publishing
West Nyack, New York 10995
Simon & Schuster, A Paramount Communications Company

PRINTED IN THE UNITED STATES OF AMERICA

DEDICATION

To Rees Lee, Principal, Adelante High School

For your long-lasting friendship, excellent sense of humor, and tireless effort to make things better for everybody.

ACKNOWLEDGMENTS

I would like to extend a hearty thanks to my students and fellow educators for putting up with the numerous riddles and puns that permeate the book.

A special thanks to Robert J. Hubbard, a talented Adelante High School student. His artwork made the pages come to life.

Finally, I'd like to thank Susan Hoehn for contributing artwork to the sections on weather and climate, astronomy and oceanography.

ABOUT THE AUTHOR

Robert G. Hoehn has taught earth science, physical science, and biology in the Roseville Union School District of California since 1963. He has received six National Summer Science grants from the National Science Foundation and has given numerous presentations to teachers and administrators attending local and state science conventions, workshops, and seminars. He has also served as a mentor teacher in his district.

Mr. Hoehn has a B.A. from San Jose State University and an administration credential from California State University, Sacramento. He is a member of the National Education Association, California Teacher's Association, and California Science Teacher's Association. Author of *Earth Science Curriculum Activities Kit,* 1991, by The Center for Applied Research in Education, he has also published a number of nonfiction books and copymaster sets and over 70 magazine articles on science education, coaching, and baseball/softball strategy.

ABOUT THIS BOOK

There are times when you need to grab and hold the attention of each student. The question is: How?

You can easily handle the problem by giving students a science starter to solve or think about. A starter activity is a short, intense assignment that may last anywhere from 15 seconds to 5 minutes.

Science Starters: Over 1,000 Ready-to-Use Attention-Grabbers That Make Science Fun for Grades 6–12, offers science educators a selection of more than 1,000 miniactivities. The three major science areas covered are physical science, earth science, and life science. Each of these areas is organized as follows:

Physical Science

- Measurement
- Matter
- Atomic Structures
- Chemistry
- Forms of Energy
- Heat and Energy
- Force and Motion
- Machines

Earth Science

- Energy Sources
- Rocks and Minerals
- Volcanoes and Earthquakes
- Fossils and Geologic Time Scale
- The Earth's Forces
- Weather and Climate
- Astronomy
- Oceanography

Life Science

- Simple Life
- Plant Life
- Animals Without Backbones
- Animals With Backbones
- Reproduction and Development
- Genetics and Change
- Human Biology
- Health and Environment

Each section includes brief, humorous activities designed to encourage students to become creative thinkers. The activities are as follows:

- What Does the Sketch Represent?
- Can You Solve the Problem?

- What's in a Name?
- Create-A-Comment
- Riddle Bits
- What Goes Where?
- Creative Potpourri

"Science Starters" are teacher-friendly because you control when and where to use them. You can challenge your students with one in the opening minutes of class or in the closing minutes of the period, or whenever you think the class needs a break from the regular routine.

Many of the activities serve as a tongue-in-cheek approach to science. This book provides numerous opportunities for teacher and student to examine the light, humorous side of science.

Special features of this book are the Teacher's Guide and Answer Key found at the end of the resource for easy reference. In many instances, there may be several responses or answers to an item. The teacher, of course, should decide what responses are appropriate.

Science Starters will supply you with a diversity of activities to promote creative thinking and stimulate learning.

TABLE OF CONTENTS

About This Book vii

PHYSICAL SCIENCE 1
    Section 1   Measurement 3
        1–1   What Does the Sketch Represent? 3
        1–2   Can You Solve the Problem? 11
        1–3   What's in a Name? 11
        1–4   Create-A-Comment 13
        1–5   Riddle Bits 15
        1–6   What Goes Where? 15
        1–7   Creative Potpourri 15
    Section 2   Matter 17
        2–1   What Does the Sketch Represent? 17
        2–2   Can You Solve the Problem? 25
        2–3   What's in a Name? 26
        2–4   Create-A-Comment 28
        2–5   Riddle Bits 30
        2–6   What Goes Where? 31
        2–7   Creative Potpourri 31
    Section 3   Atomic Structures 32
        3–1   What Does the Sketch Represent? 32
        3–2   Can You Solve the Problem? 40
        3–3   What's in a Name? 41
        3–4   Create-A-Comment 42
        3–5   Riddle Bits 44
        3–6   What Goes Where? 44
        3–7   Creative Potpourri 45
    Section 4   Chemistry 47
        4–1   What Does the Sketch Represent? 47
        4–2   Can You Solve the Problem? 55
        4–3   What's in a Name? 55
        4–4   Create-A-Comment 57
        4–5   Riddle Bits 59
        4–6   What Goes Where? 60
        4–7   Creative Potpourri 61

Section 5 Forms of Energy 62
- 5–1 What Does the Sketch Represent? 62
- 5–2 Can You Solve the Problem? 70
- 5–3 What's in a Name? 70
- 5–4 Create-A-Comment 71
- 5–5 Riddle Bits 73
- 5–6 What Goes Where? 73
- 5–7 Creative Potpourri 74

Section 6 Heat and Energy 75
- 6–1 What Does the Sketch Represent? 75
- 6–2 Can You Solve the Problem? 83
- 6–3 What's in a Name? 83
- 6–4 Create-A-Comment 84
- 6–5 Riddle Bits 86
- 6–6 What Goes Where? 86
- 6–7 Creative Potpourri 87

Section 7 Force and Motion 88
- 7–1 What Does the Sketch Represent? 88
- 7–2 Can You Solve the Problem? 96
- 7–3 What's in a Name? 97
- 7–4 Create-A-Comment 98
- 7–5 Riddle Bits 100
- 7–6 What Goes Where? 100
- 7–7 Creative Potpourri 101

Section 8 Machines 102
- 8–1 What Does the Sketch Represent? 102
- 8–2 Can You Solve the Problem? 110
- 8–3 What's in a Name? 110
- 8–4 Create-A-Comment 111
- 8–5 Riddle Bits 113
- 8–6 What Goes Where? 113
- 8–7 Creative Potpourri 114

EARTH SCIENCE 115

Section 9 Energy Sources 117
- 9–1 What Does the Sketch Represent? 117
- 9–2 Can You Solve the Problem? 123
- 9–3 What's in a Name? 124
- 9–4 Create-A-Comment 126
- 9–5 Riddle Bits 128
- 9–6 What Goes Where? 128
- 9–7 Creative Potpourri 129

Section 10 Rocks and Minerals 131
- 10–1 What Does the Sketch Represent? 131
- 10–2 Can You Solve the Problem? 139

Table of Contents   xi

    10–3   What's in a Name?   139
    10–4   Create-A-Comment   143
    10–5   Riddle Bits   145
    10–6   What Goes Where?   145
    10–7   Creative Potpourri   146

Section 11   Volcanoes and Earthquakes   147
    11–1   What Does the Sketch Represent?   147
    11–2   Can You Solve the Problem?   154
    11–3   What's in a Name?   155
    11–4   Create-A-Comment   157
    11–5   Riddle Bits   159
    11–6   What Goes Where?   159
    11–7   Creative Potpourri   160

Section 12   Fossils and Geologic Time Scale   161
    12–1   What Does the Sketch Represent?   161
    12–2   Can You Solve the Problem?   169
    12–3   What's in a Name?   170
    12–4   Create-A-Comment   171
    12–5   Riddle Bits   173
    12–6   What Goes Where?   174
    12–7   Creative Potpourri   174

Section 13   The Earth's Forces   176
    13–1   What Does the Sketch Represent?   176
    13–2   Can You Solve the Problem?   183
    13–3   What's in a Name?   184
    13–4   Create-A-Comment   186
    13–5   Riddle Bits   188
    13–6   What Goes Where?   189
    13–7   Creative Potpourri   189

Section 14   Weather and Climate   191
    14–1   What Does the Sketch Represent?   191
    14–2   Can You Solve the Problem?   196
    14–3   What's in a Name?   196
    14–4   Create-A-Comment   199
    14–5   Riddle Bits   200
    14–6   What Goes Where?   201
    14–7   Creative Potpourri   202

Section 15   Astronomy   204
    15–1   What Does the Sketch Represent?   204
    15–2   Can You Solve the Problem?   207
    15–3   What's in a Name?   208
    15–4   Create-A-Comment   210
    15–5   Riddle Bits   212
    15–6   What Goes Where?   212
    15–7   Creative Potpourri   213

Section 16 Oceanography 215
- 16–1 What Does the Sketch Represent? 215
- 16–2 Can You Solve the Problem? 222
- 16–3 What's in a Name? 224
- 16–4 Create-A-Comment 226
- 16–5 Riddle Bits 228
- 16–6 What Goes Where? 229
- 16–7 Creative Potpourri 229

LIFE SCIENCE 231

Section 17 Simple Life 233
- 17–1 What Does the Sketch Represent? 233
- 17–2 Can You Solve the Problem? 241
- 17–3 What's in a Name? 242
- 17–4 Create-A-Comment 243
- 17–5 Riddle Bits 245
- 17–6 What Goes Where? 245
- 17–7 Creative Potpourri 246

Section 18 Plant Life 248
- 18–1 What Does the Sketch Represent? 248
- 18–2 Can You Solve the Problem? 255
- 18–3 What's in a Name? 256
- 18–4 Create-A-Comment 258
- 18–5 Riddle Bits 260
- 18–6 What Goes Where? 260
- 18–7 Creative Potpourri 261

Section 19 Animals Without Backbones 262
- 19–1 What Does the Sketch Represent? 262
- 19–2 Can You Solve the Problem? 270
- 19–3 What's in a Name? 271
- 19–4 Create-A-Comment 273
- 19–5 Riddle Bits 275
- 19–6 What Goes Where? 275
- 19–7 Creative Potpourri 276

Section 20 Animals With Backbones 278
- 20–1 What Does the Sketch Represent? 278
- 20–2 Can You Solve the Problem? 284
- 20–3 What's in a Name? 284
- 20–4 Create-A-Comment 287
- 20–5 Riddle Bits 289
- 20–6 What Goes Where? 289
- 20–7 Creative Potpourri 290

Section 21 Reproduction and Development 291
- 21–1 What Does the Sketch Represent? 291
- 21–2 Can You Solve the Problem? 297

Table of Contents   xiii

    21–3   What's in a Name?   298
    21–4   Create-A-Comment   301
    21–5   Riddle Bits   303
    21–6   What Goes Where?   304
    21–7   Creative Potpourri   304

Section 22   Genetics and Change   306
    22–1   What Does the Sketch Represent?   306
    22–2   Can You Solve the Problem?   314
    22–3   What's in a Name?   315
    22–4   Create-A-Comment   316
    22–5   Riddle Bits   318
    22–6   What Goes Where?   318
    22–7   Creative Potpourri   319

Section 23   Human Biology   321
    23–1   What Does the Sketch Represent?   321
    23–2   Can You Solve the Problem?   329
    23–3   What's in a Name?   331
    23–4   Create-A-Comment   333
    23–5   Riddle Bits   335
    23–6   What Goes Where?   335
    23–7   Creative Potpourri   335

Section 24   Health and Environment   337
    24–1   What Does the Sketch Represent?   337
    24–2   Can You Solve the Problem?   345
    24–3   What's in a Name?   346
    24–4   Create-A-Comment   348
    24–5   Riddle Bits   350
    24–6   What Goes Where?   350
    24–7   Creative Potpourri   351

TEACHER'S GUIDE AND ANSWER KEY   353
PHYSICAL SCIENCE   353
    Section 1   Measurement   353
    Section 2   Matter   355
    Section 3   Atomic Structures   357
    Section 4   Chemistry   360
    Section 5   Forms of Energy   363
    Section 6   Heat and Energy   365
    Section 7   Force and Motion   367
    Section 8   Machines   370

EARTH SCIENCE   373
    Section 9   Energy Sources   373
    Section 10   Rocks and Minerals   376
    Section 11   Volcanoes and Earthquakes   379
    Section 12   Fossils and Geologic Time Scale   381

Section 13	The Earth's Forces	384
Section 14	Weather and Climate	387
Section 15	Astronomy	390
Section 16	Oceanography	393

LIFE SCIENCE 396

Section 17	Simple Life	396
Section 18	Plant Life	398
Section 19	Animals Without Backbones	400
Section 20	Animals with Backbones	403
Section 21	Reproduction and Development	406
Section 22	Genetics and Change	408
Section 23	Human Biology	411
Section 24	Health and Environment	413

PHYSICAL SCIENCE

- ▼ MEASUREMENT
- ▼ MATTER
- ▼ ATOMIC STRUCTURES
- ▼ CHEMISTRY
- ▼ FORMS OF ENERGY
- ▼ HEAT AND ENERGY
- ▼ FORCE AND MOTION
- ▼ MACHINES

Section 1

MEASUREMENT

1-1 What Does the Sketch Represent?

Let students write on the line below each sketch what they think the illustration represents. Ask them to match their descriptions with the theme of the section.

1.

2.

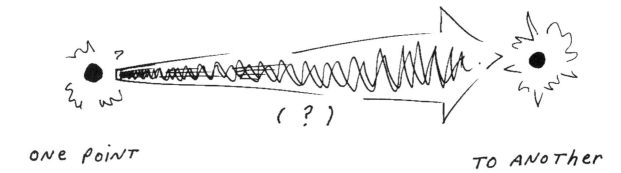

ONE POINT (?) TO ANOTHER

3.

4.

5.

6.

7.

8.

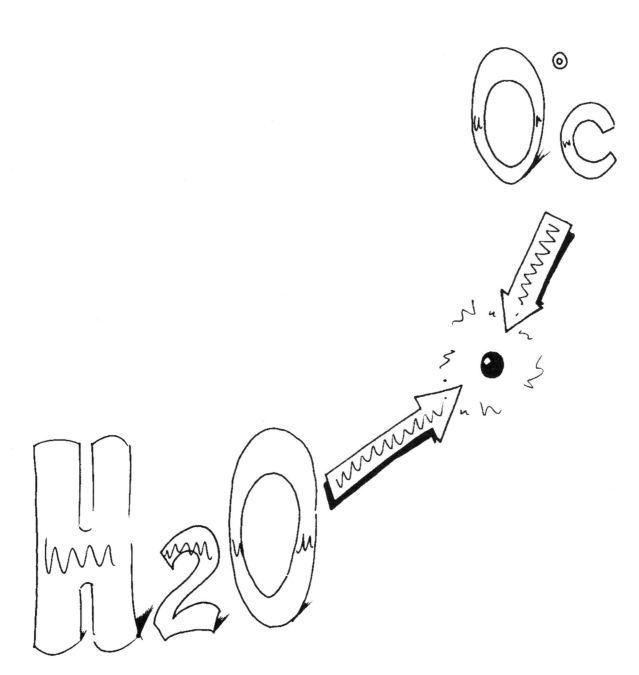

Physical Science

MEASUREMENT (continued)

1–2 Can You Solve the Problem?

These six problems will be a measure of student stamina. It will also test their ability to be flexible thinkers.

1. One inch equals 2.54 centimeters. Thus a centimeter is this long:

   What can be found in the middle of a centimeter?
2. Mass is the amount of matter in an object. List *two* words that have *mass* as the first part of the word and list *two* words that have *mass* as the last part of the word.
3. Rosemont's animal shelter has 23 dogs and 15 cats. How many *feet* are in a *pound*?
4. How many millimeters are in an inchworm?
5. There are 1,000 milliliters in one liter. Show how to make *one liter* out of 999 milliliters.
6. It is extremely important to make careful measurements when handling chemicals in an experiment. Circle the letters in a term below that completes the following sentence:

   Accuracy counts because you want to be _____ .

   millimeter
   measure
   experiment
   metric
   balance
   exponent

1–3 What's in a Name?

These five activities require students to create names of people, places, or things from measurement terms.

1. Create three fictitious rock band names from the terms and prefixes. Combine two or more terms.

   Example: milli (prefix) and meter
   Milli Meter and the Metrics

   centi   kilo
   gram   mass
   liter   celsius
   deci   ruler
   meter   degree

MEASUREMENT (continued)

2. Write the names of five items that have *gram* as part of their names. *Note:* You can't use any part of the metric system as names.
3. See if you can name *three* animals with three letters in their names from the letters in VOLUME. You may use a letter more than once.
4. As you read the message, cross out the letters in

    CIRCUMFERENCE

    "The third letters of the alphabet should go. Take the *e*'s out too! Say goodbye to the *i*. Go ahead and cross out the *r*'s. Now remove the *m*."

    If you did everything correctly, the remaining three letters (unscrambled) will spell out what this activity should have been.

    This activity should have been _____.

5. See if you can write the names of eight different animals using the letters in CENTIGRAM. You may use a letter more than once.

MEASUREMENT (continued)

1–4 Create-A-Comment

Study the two illustrations. Then create a statement or comment that matches the theme of the section. Make it as humorous as you want. Place the comment in the balloon next to the figure.

10.

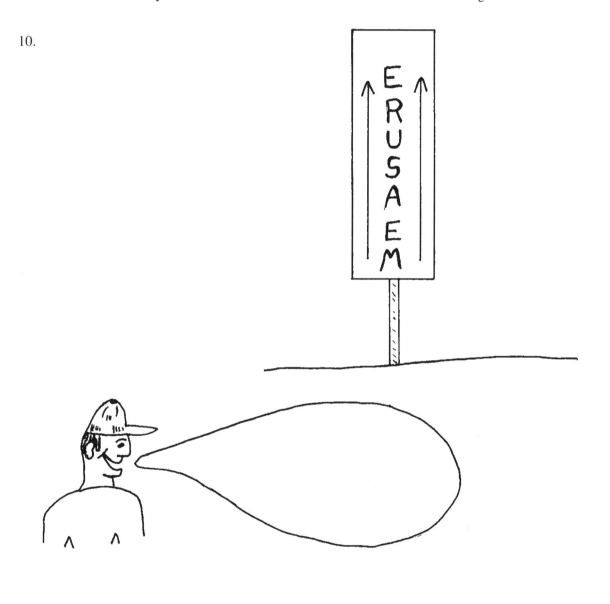

11.

Physical Science   15

MEASUREMENT (continued)

1–5 Riddle Bits

Riddles serve as a minichallenge for creative thinkers. Here are ten stimuli for anxious students.

1. Even the domestic animals in Europe are familiar with the metric system. Why are cats more aware than any other animal?
2. How many meters fit in a yard?
3. What animal's name makes up three-fourths of a unit of measurement in the metric system?
4. What animal's name makes up three-fourths of a unit of measurement?
5. A certain prefix of measurement is sick all the time. Which one do you think it is?
6. What does the sketch represent?

$$1"\quad INCH \quad 1"$$

7. Why do dieters prefer their liquids in liters?
8. What liquid measurement is the most creative?
9. What measurement has $37\frac{1}{2}$ percent of its name made up of an object used for fuel?
10. What measurement spells the name of a face when the letters are unscrambled?

1–6 What Goes Where?

Have students identify words or phrases and symbols. Then ask them to arrange everything in its correct order to reveal the mystery term.

1. "grrr" + sounds like "ease" + e + d.
2. Sounds like "kite" + means sicklike + m + r + i.
3. Sounds like "amateur" + one small cube (–a).
4. er + "you and _____" + sounds like [key] + "Not high, but _____" + t.
5. g + [eye] + penny + [ram] sheep.
6. Alcoholic beverage + c + s.

1–7 Creative Potpourri

These six miniproblems engage students in active, creative thinking.

1. Janice measured her lab table. It was 80 centimeters wide and 320 centimeters long. Show a

MEASUREMENT (continued)

novel way to express the *area* of her table in square centimeters. *Note:* Do not write 25,600 cm² for an answer.

2. How could you create a spring scale from the word *scale*?
3. Look at the list of letters strung together. How many METERS can you find?

MEMETERMETERMETERMETERMETERMETERMETERMETERMETERMETERMET

4. Show, using a simple sketch, how *three* meters can fit in *one* yard.
5. Use the letters in *metric system* to form chemical symbols. You may use a letter only once with the exception of letters c, e, i, t, m, r, and s. These letters may be used twice. *Note:* Let your creative genie be your guide.
6. As you know, mass is a measure of the amount of matter. Degrees are divisions or units of a scale, as of a thermometer. Use the letters in MASS to indicate two different *degrees*.

Section 2

MATTER

2–1 What Does the Sketch Represent?

The eight activities in this section allow students to select a sketch and describe what it represents. They have freedom to use humor in matching a description with the section theme.

1.

2.

_____

3.

4.

5.

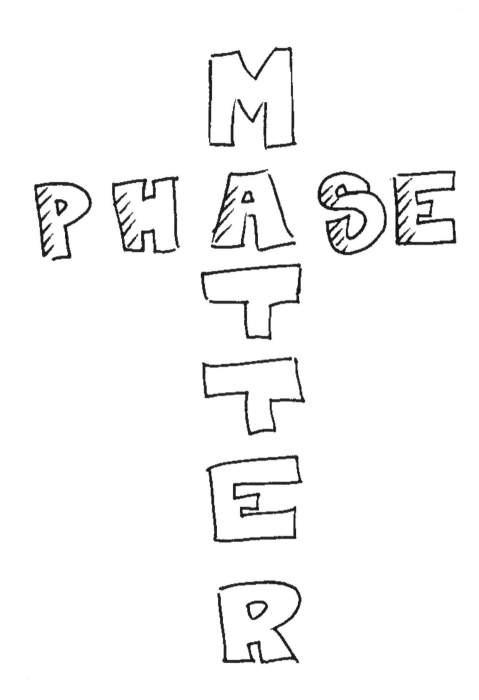

6.

7.

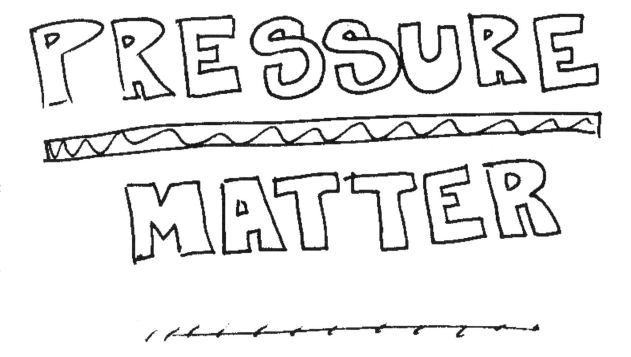

8.

Physical Science

MATTER (continued)

2–2 Can You Solve the Problem?

Miniproblem activities rev up the neurons and coax students out of the starting blocks.

The following five activities help students concentrate on anything that has weight and occupies space—in a word, matter.

1. The box contains two-, three-, and four-letter combinations. Combine *six* of them to spell the names of *three physical properties* of matter.

   ```
              den         or      hard
   ubi              mag       sol
                               sub
          col   sium      lity
                               sity
          ne    ness      cal
   ```

2. The circle contains two-, three-, four-, and five-letter combinations. Combine *six* of them to spell the names of *three chemical properties* of matter.

   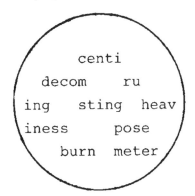

3. Each word is an example of matter. Within each word is *another* example of matter. Find the matter within matter. However, you *can't* move letters around.

   Example: hair - air

   | tarsus | feldspar | whelk | saloon | shrimp |
   | monkey | squash | pupa | stalagmite | travertine |

4. Unscramble the words and reveal the example of matter. You need to unscramble the first part or last part of the word.

MATTER (continued)

Example: ylaptpus - platypus

spothtgil	nelacdar
luvture	gargeab
rotdne	emsquite
saugase	merycru
lisver	orfceps

5. List ten examples of matter using human body structures as part of the examples.

Example: ear - bear

nail	organ
rib	hip
cell	gum
hair	liver
arm	iris

2–3 What's in a Name?

These six activities let students create names of people, places, or things from various kinds of matter.

1. Rocks, water, soil, and minerals cover the Earth's crust. Use the letters in SOIL and WATER to spell the names of two items usually found in a garage. You may use a letter only once.

2. Use the letters in MATTER as the first letters of *six* items found in some closets. The first item begins with "m," the second with "a," and so on.

3. Another word for matter is STUFF. Use the letters in STUFF as the first letters of *five* items found in or on many automobiles. The first item begins with "s," the second, with "t," and so on.

4. There are two or three words listed for each numbered item. If you put the words in their correct order (and use your imagination), you will reveal the name of living matter known to inhabit the Earth.

Example: key + on = onkey or *monkey*.

a. ill + arm + ad	e. poor + whip
b. or + at + all	f. pod + halo
c. am + pop + hip	g. get + ion + at
d. ant + tar	h. ad + raw

5. Find the mystery animal. There are *five* sets of two-letter combinations in the box. Put them together correctly and reveal the answer. Here are three hints:

MATTER (continued)

- Oldest known bird
- About the size of a crow
- An evolutionary link from reptiles to birds

| yx | ch | pt | eo | ar |

Wait a minute! A letter is missing! What is it?

6. Matter is anything that has weight and occupies space. Use the letters in ANYTHING (omit the *Y*) as the first letters of *seven* minerals found in the Earth's crust. The first mineral begins with "a;" the second with "n;" and so on.

MATTER (continued)

2–4 Create-a-Comment

This activity allows students to create humor to match the section theme.

Provide students with a copy of a drawing. Have them write a comment or statement in the balloon next to the figure in the illustration.

1.

2.

MATTER (continued)

2–5 Riddle Bits

Here are ten riddles for students to tackle. Riddles test a student's ability to be a flexible thinker while emitting a groan or two.

1. What did Mrs. Matter say about her 2-year-old daughter, Donna, when Donna threw a temper tantrum?
2. What number is hidden in the word MATTER?
3. What mineral appears in every chemical change?
4. How many elements are in a CHeMICaL CHANGe?
5. If you shake a can of soda pop, what kind of change occurs?
6. What does the following represent?

   $\underline{m} - 2" - \underline{a} - 2" - \underline{t} - 1" - \underline{t} - 3" - \underline{e} - 2" - \underline{r}$

7. MA(TT)ER
   Why are the two T's in MATTER circled?
8. How can a liquid be changed to a solid?
9. How much do these two items for matter weigh? $2n + 2n$
10. What does the sketch illustrate?

Physical Science

MATTER (continued)

2–6 What Goes Where?

These activities combine numbers, symbols, letters, and phrases to identify a term related to matter. All items must be arranged correctly to reveal the term.

1. ol + mmm + "you" + v.
2. t + lion's home + silicon (chemical symbol) + y.
3. Two words
   First word: (.) + ic.
   Second word:

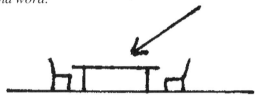

4. s + m + 16 ounces + co.
5. e + t + opposite of women + Spanish word for *the*.
6. les + copper (chemical symbol) + an underground animal.

2–7 Creative Potpourri

This section provides a minicollection of activities. Once again, these make excellent openers and closers.

1. A TURKEY is matter because it occupies space and has weight. The word KEY in TURKEY is also matter. See if you can find three more examples of matter in TURKEY.
2. Use a straight line to show a way MATTER can undergo a physical change.
3. Burning is an example of a chemical change. Use a straight line to show how BURNING may go through a physical change.
4. How many different kinds of matter make up MAGNESIUM?
5. Consider a mixture of iron and sand. You can separate the mixture by recovering iron particles with a magnet. Show how a mixture can be separated without using magnets, heat, water—or any special device or procedure.

Section 3

Atomic Structures

3–1 What Does the Sketch Represent?

Flexible thinking takes center stage in this activity. Students coax their imaginative powers to spring into action.

Ask students to study the sketch. Then have them write what they think the illustration represents on the line below the drawing.

1.

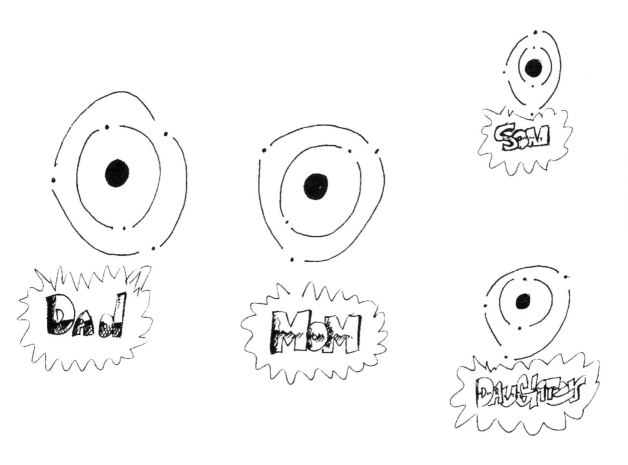

2.

_____

3.

_____

4.

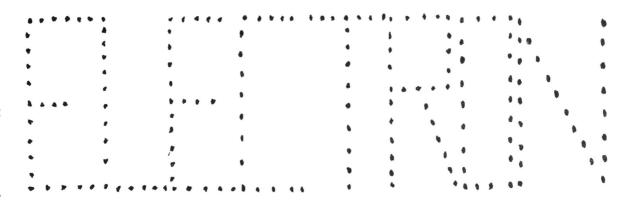

5.

6.

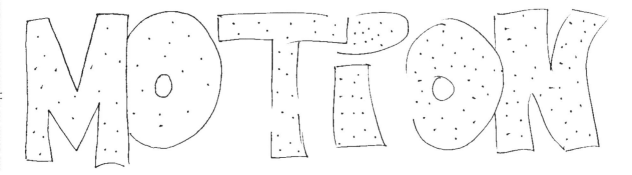

7.

8.

ATOMIC STRUCTURES (continued)

3–2 Can You Solve the Problem?

This section offers five miniproblems to help students warm up their thinking engines.

1. Use the letters in NUCLEUS to identify land or sea animals found on Earth. Use each letter, in turn, to be the starting letter for the animal's name. For example, *N* would be the first letter for NEWT.
2. Use the letters in PROTON to identify minerals found in the Earth's crust. Use each letter, in turn, to be the starting letter for the name of the mineral. For example, *P* would be the first letter for PYRITE.
3. Use the letters in ELECTRON to identify items found in many homes. Use each letter, in turn, to be the starting letter for the name of the item. For example, *E* would be the first letter in ELECTRIC STOVE.
4. Let the hints in the right-hand column help you find the missing letters in the left-hand column. When you've completed the puzzle, answer the question: What is a NEUTRON?

	Missing Letters	*Hints*
a.	__ N __	a. Small social insect
b.	__ __ __ E __ __	b. Cut into two equal parts
c.	__ U __ __	c. Onion-shaped root
d.	__ __ __ T __	d. Needed for chewing
e.	__ __ R __ __	e. To swear
f.	__ O __ __ __ __	f. False hair
g.	__ __ N __ __ __ __	g. Mass of matter per unit of volume

5. The answer to the question below the puzzle appears twice in the puzzle. Shade in the letters that spell the answer. The letters to the answer may be found up, down, horizontal, and backward. The shaded letters will give you the answer to this question:

    What is the electrical charge of a proton?

t	o	n	p	p	r	o
p	r	n	r	s	o	r
t	n	s	o	o	r	p
p	r	o	t	o	n	s
s	t	o	o	p	r	n
r	n	s	n	t	p	o
p	t	o	s	r	o	n

    What subatomic particles beginning with the letter P are found in the nucleus of an atom?

Physical Science

ATOMIC STRUCTURES (continued)

3–3 What's in a Name?

Students will be scratching their heads at a steady pace searching for just the right name. These five activities should provide a comfortable challenge.

1. Try this toughie. Find a name that rhymes with each atomic term in the left-hand column. The clues are listed in the right-hand column. Place answers in the middle column.

Atomic Terms	Rhyming Name	Clues
a. electron	a.	a. Now called Ho Chi Minh City
b. proton	b.	b. A sugar-plum
c. energy	c.	c. Apathy
d. bond	d.	d. Leaf of a fern or palm
e. shell	e.	e. To inhabit

2. Remove the letters RON from ELECTRONS. Now use the letters in ELECTS to spell the words that will complete the following sentences. A letter may be used more than once.

   a. Ron should ___ ___ ___ Connie drive his car.
   b. Ron has a pet ___ ___ ___ in his fish tank.
   c. Connie has three dollars ___ ___ ___ ___ than Ron.
   d. Ron likes to ___ ___ ___ ___ ___ ___ his own clothes.
   e. John wanted Ron to ___ ___ ___ ___ his stereo.
   f. Ron walked 3 miles in the ___ ___ ___ ___ ___ .

3. See if you can make four boys' names out of the letters in ATOMIC. A letter may be used more than once.

4. List the names of four vegetables using the letters in NUCLEAR ENERGY POWER to create the words. A letter may be used more than once.

5. List the names of ten fruits using the letters in PERIODIC TABLE OF ELEMENTS to create the words. A letter may be used more than once.

42 Physical Science

ATOMIC STRUCTURES (continued)

3–4 Create-A-Comment

This section provides two cartoon activities that allow students to show their humorous side.

Begin by giving students one or both drawings. Ask them to examine the sketch and write a comment or statement in the balloon above the two figures in the illustration.

1.

© 1993 by The Center for Applied Research in Education

2.

ATOMIC STRUCTURES (continued)

3–5 Riddle Bits

One way to change the pace and elicit groans is by offering students riddles to solve. The following 12 riddles should do the trick.

1. Where might "atomic weather" originate?
2. What do these numbers represent? 1 – 20 – 15 – 13 – 9 – 3
3. Where can you find underwater electron particles?
4. What is a nuclear scientist's favorite piece of furniture?
5. How can you get two negative charges from a positive particle?
6. What fruit often used as a vegetable, can be spelled by using the letters in ATOM? (Two of the letters may be used twice.)
7. How many ATOMS are in a million pencil dots?
8. If an atom's electrons are in "tibro," what can be said about them?
9. If an atom could be an animal, what animal might it be?
10. What would you get if you combined the ends of NEUTRON?
11. What university did most nuclear scientists attend?
12. What animal is made up of cobalt and tungsten atoms?

3–6 What Goes Where?

Offer students the following mixed symbols, words, numbers, and phrases to organize. If they put everything together correctly, they will identify the term.

1. opposite of off + t + "gnu" + r.
2. [1000 lbs]—[1000 lbs] + not an amateur (abbr.) (three letters).
3. n + [eye] + 0
4. Two words

    First word: y + 60 percent of NERVE + g + e.

    Second word: 1 + not he, but __ __ __ + 1.

Physical Science

ATOMIC STRUCTURES (continued)

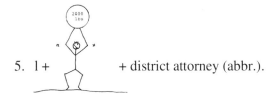

5. 1 + [figure] + district attorney (abbr.).

6. Two words

   First word: two-thirds of a NUT + 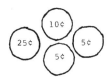 + cl.

   Second word:

7. Two words

   First word: [picture] (−s).

   Second word: __ __ __ and eggs + c + cucum __ __ __.

3–7 Creative Potpourri

Students experience creative freedom in these activities. Their imaginations do strange things with words, sentences, and puzzles. The activities are fun to do and test the ability of a student to be a flexible thinker.

1. Make silly sentences out of the following terms. However, before you do, you must separate the terms into two words. For example, *radioactive* might look like this:

   It seems like Tom's *active* only when he plays the *radio*.

electron	radioisotope
alchemist	copper
proton	plumbum

2. Create graffiti related to atomic structures or anything dealing with nuclear power. For example,

ATOMIC STRUCTURES (continued)

ATOMS ARE RAD

3. The Dalton Atomic Model suggests that matter is made up of atoms. Think of a way to show this may be false.
4. List *five* words by using the letters in ATOMS that relate in some way to science. Put each word in a sentence (be creative) to show the relationship. You may use a letter more than once. You do not have to use all the letters. Also, the words must contain four or more letters.
5. See how many recognizable words you can create by combining two or more chemical symbols. For example,

   carbon – C + argon – Ar = CAr or *car*

Section 4

CHEMISTRY

4–1 What Does the Sketch Represent?

This activity provides students with eight sketches to examine and describe.

Offer students one sketch at a time. Have them match a description with the section theme. Encourage students to use creative expression, have fun, and write what they think the illustration represents on the line below the sketch.

1.

"nitrogen ... nitrogen gas ... nitrogen ... atomic number 7 ... liquid nitrogen ... nitrogen monoxide ..."

47

2.

3.

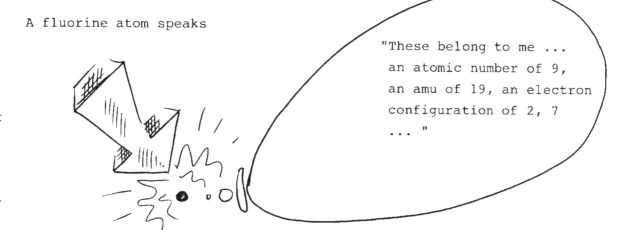

4.

5.

MOLE(CCCC)CULES

_____

6.

dot or dots

7.

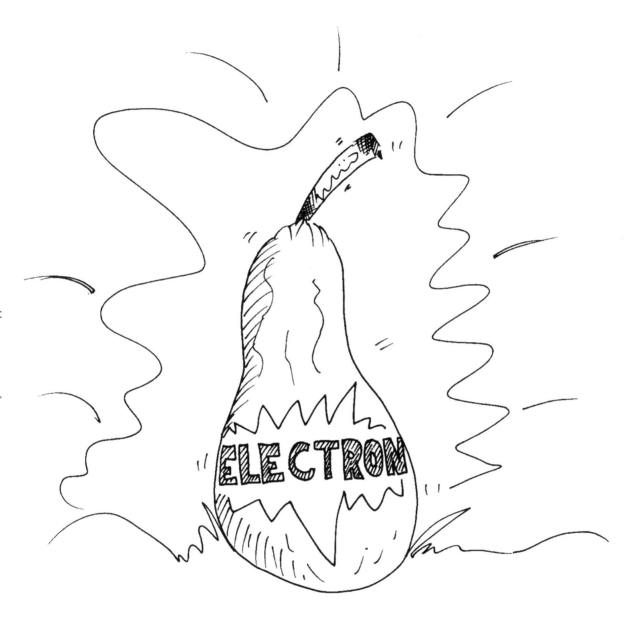

8.

Physical Science

CHEMISTRY (continued)

4–2 Can You Solve the Problem?

Here are five miniproblems to put students in a chemistry mode (or at least point them in the general direction.)

1. Interesting things happen when certain substances combine. If you *correctly* combine the substances in the box, what might occur?

Rb (–b) + Th (–h) + Y + I + S + Mo (–o) + H + C + Er (–r)

2. Hydrogen, the simplest element, is a gas at ordinary temperatures. Consider the word *IT*. Funk & Wagnall's dictionary states that *IT* represents some implied idea, condition, action, or situation. So let's take action. How can you get hydrogen from *IT*?

3. Six properties of water appear in the box. However, a problem exists. What problem? The words identifying the physical properties are scrambled, plus there are three extra words that have nothing to do with the physical properties of water. Try to identify the six physical properties of water.

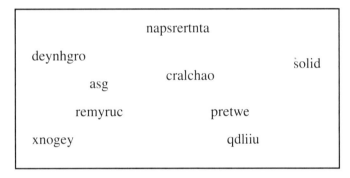

4. A diamond is a mineral consisting of carbon crystallized under great pressure and temperature. Think of a way to make a diamond out of *four* carbons.

5. Chemistry deals with the composition of substances and the changes in composition which these substances undergo. Use the letters in *chemistry* as the first letters of chemical substances found on Earth. Identify *two* examples for each letter.

4–3 What's in a Name?

Students should enjoy creating names from or locating names within chemistry terms. This section offers students five activities guaranteed to push the creative button.

1. Make up rhyming first names of boys and/or girls from the chemistry terms. You may abbreviate names; however, do not use nicknames. The first one is done for you.

CHEMISTRY (continued)

Example: methyl - Ethyl Methyl

ion	base	alloy
calorie	meter	mole
salt	weight	lye

2. Make up first and last names for girls from the chemistry terms. You may abbreviate names; however, do not use nicknames. The first one is done for you.
   Example: element - Ella Mint

energy	generator
polymer	evaporation
molecule	deliquescence
milliliter	

3. List *five* chemistry terms that contain the name of an animal.
   Example: gram - ram

4. List *five* chemistry terms that contain the names of plants or names of items related to plants.
   Example: <u>sod</u>ium - sod

5. List *five* chemistry terms that contain meteorology (study of weather) names.
   Example: che<u>mist</u>y - mist

Physical Science 57

CHEMISTRY (continued)

4–4 Create-A-Comment

These two activities allow students to create humorous comments or statements to match the theme of the sketch. There are no right or wrong answers.

Give students a copy of a drawing. Ask them to write a comment or statement in the balloon next to the figure in the illustration.

1.

2.

Physical Science

CHEMISTRY (continued)

4–5 Riddle Bits

Here's where the creative keg gets a strong tap. Offer several riddles to students. This will test their aptitude for flexible thinking and their strength for controlling the urge to kill.

1. What do you call this list of names?

    hereford
    guernsey
    holstein
    black angus

2. What two atomic numbers on the Periodic Table of Elements are always late?
3. What atomic numbers of the Periodic Table of Elements combine for the purpose of amusement?
4. What rare earth element is Santa Claus's favorite? (He uses it repeatedly.)
5. What element is covered half-way with a soft, fine, thick animal hair?
6. What common insect makes up about $37\frac{1}{2}$ percent of a nonmetal element?
7. Concentrated sulfuric acid is also known as oil of vitriol. Therefore, where might you find the oil?
8. Student A wore a T-shirt like this:

    Student B wore a T-shirt like this?

    Which student, A or B, would be a good friend?
9. Funk & Wagnall's refers to a scientific law as a formal statement of the manner or order in which a set of natural phenomena occur under certain conditions.

    What *law* might occur if these items combined and mixed together?

    cabbage   bell pepper
    onion   sugar
    mayonnaise   vinegar
    milk

CHEMISTRY (continued)

10. H_2SO_4 is the chemical formula for sulfuric acid. SiO_2 is the chemical formula for silicon dioxide. What is T_4?

11. Ron wrote a report on *ionization*. He did a rotten job of researching the subject. What *word*, to summarize Ron's effort, did Mr. Bromine, Ron's chemistry teacher, write at the top of the paper?

12. The wheels on Jean's wagon rusted and would barely turn. He asked his mother to pour boiling water over them. Why do you think Jean made this request?

13. How is the sketch related to chemistry?

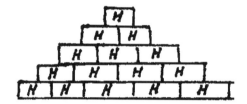

14. Matter occurs in several states. It may appear as a solid, liquid, or gas. In what state might you find uranium, tantalum, and hydrogen?

15. In how many states would you find matter?

4–6 What Goes Where?

Have students decipher symbols and phrases. Then ask them to organize everything to reveal the correct term.

     B+
1. or + i + penny.
     C
2. "Fuss" + opposite of less + a.
3. Iron, for example, + id + ? an behold.
4. A + T^2 + E + M + R.
5. Sounds like "meow" + 4 + ? Vegas, Nevada.
6. id + sounds like "coal" + o.
7. "bull" + ci + prefix for badly or incorrectly.
8. dr + "8" + hydrogen (first two letters).
9. "your" + "eight" + at + 19th letter of the alphabet.
10. "ur" + or + fate (−f and e) + sounds like "gin."

Physical Science

CHEMISTRY (continued)

4–7 Creative Potpourri

The following four activities are a mixture of zany definitions, strange sentences, and gamelike, thought-provoking exercises.

1. A mixture is a material consisting of two or more kinds of matter, each retaining its own characteristic properties. For example, radishes and carrots. How many mixture combinations are in the box? Name them.

gravel	sand
NaCl	Fe filings
S	confetti

2. Make up zany definitions for the chemistry terms. The first one is done for you.

   Example: element - a new brand of candy mint

amalgam	lignite
ketones	volatile
iodine	sulfate

3. Use the chemistry terms to write strange sentences. For example, the word *amorphous* might look like this: "Okay, don't eat your vegetables. That'll leave *amorphous* (more for us)."

mixture	fungicide
metal	hydride
monatomic	dynamite

4. Make up two or three personalized car license plates or bumper stickers related to chemistry terms.

   Examples:

   License plate: ORG CHMST   (organic chemist)
   Bumper sticker: Chemists stir things up.

Section 5

FORMS OF ENERGY

5–1 What Does the Sketch Represent?

The eight illustrations offer a fun, relaxing way for students to think about a subject. So have them examine each sketch and write what they think the drawing represents on the line below the sketch.

1.

2.

3.

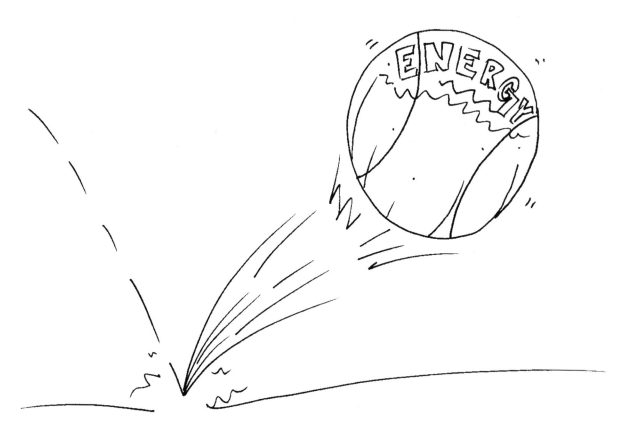

4.

5.

6.

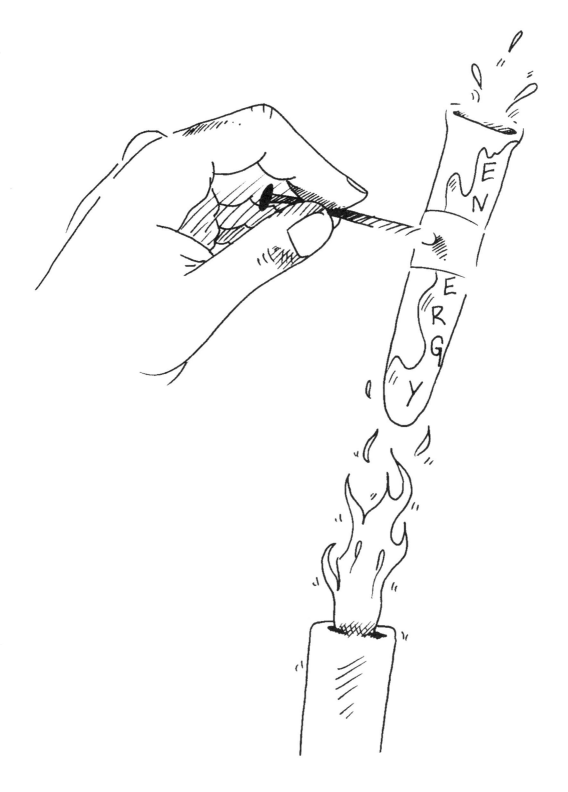

7.

8.

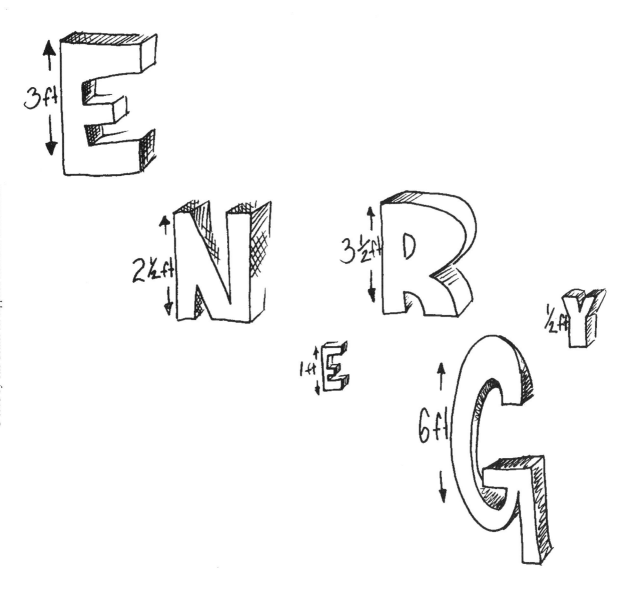

FORMS OF ENERGY (continued)

5–2 Can You Solve the Problem?

These five activities will help students put their creative energy to work.

1. Jane bungee jumped off a construction crane. Barbara, Jane's friend, said, "Jane, you've got lots of NERVE." How much ENERGY was needed to build up Jane's NERVE?
2. Use the letters in HEAT ENERGY to spell the names of three items: a drink, a domestic animal, and a wild animal. You may use the same letter more than once.
3. Show how you can get two-thirds ENERGY out of ELECTRICITY.
4. Our eyes receive light energy. The brain interprets this energy as shapes, colors, relative brightness, and distance. Sketch an eye that allows the brain to do this. Draw a second one that doesn't allow the brain to interpret light energy.
5. Devise a simple way of producing mechanical energy by using any three of the following items: block of wax, string, hammer, mud, wooden board, water, and a nail.

5–3 What's in a Name?

These five activities allow students to create names of plants, animals, rocks/minerals, and cartoon characters from the letters in the terms.

1. Use the letters in NUCLEAR POWER PLANT to form the names of three green vegetables. You may use letters more than once.
2. Use the letters in KINETIC ENERGY to form the names of two insects. You may use letters more than once.
3. Use the letters in POTENTIAL ENERGY to form the names of six different minerals. You may use letters more than once.
4. Use the letters in MAGNETIC POWER to form the names of three bird cartoon characters. You may use letters more than once.
5. Use the letters in LIGHT AND SOUND ENERGY to form the names of six different rocks. You may use letters more than once.

FORMS OF ENERGY (continued)

5–4 Create-A-Comment

Give students the two illustrations in this activity to examine. Have them look over each copy and think about the theme of the section: Forms of Energy.

Then ask them to write a comment or statement in the balloon to the side of the figure.

1.

2.

Physical Science

FORMS OF ENERGY (continued)

5–5 Riddle Bits

Here are 12 riddles for students to moan over. They should experience a creative challenge for several minutes.

1. What part of ENERGY shows the work done in moving a body one centimeter against the force of one dyne?
2. POTENTIAL ENERGY has 15 letters. If you removed 10, how many would be left?
3. How much SOUND can you get from VIBRATIONS?
4. What two metals are part of MAgNETiC ENeRGY?
5. What form of energy contains a human structure?
6. What form of energy do people consume the most?
7. What would you get if you combine SOUND and LIGHT?
8. Why does 44 percent of potential energy stay dry?
9. Ric Jones won the election for Student Body President. His campaign was packed with energy. Where did the energy come from?
10. What holds KINETIC energy together?
11. Let's say you write the words SOLAR ENERGY on paper with a pencil. If you examine the letters carefully, you'll be able to answer this question: What holds SOLAR ENERGY together?
12. How can you show energy on the move?

5–6 What Goes Where?

Students will stay focused while they unscramble symbols, words, and numbers to identify the correct terms.

1. Two words
   First word: tee (–ee) + 8 (–e, i and t) + lie (–e).
   Second word:

2. Two words
   First word: s + core (–c) + d + t.
   Second word: sounds like "rrr" + n + sounds like "gee."

FORMS OF ENERGY (continued)

3. d + three-fifths of an ounce + 19th letter.
4. frozen water (−e) + male turkey + first letter in alphabet.
5. c + sounds like the opposite of old + 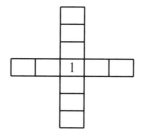 + 1.
6. a + 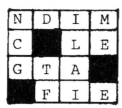 + id (spelled backward) + r.

5–7. Creative Potpourri

These five items represent a mixture of various activities. These activities should awaken sleeping neurons.

1. What two forms of energy crossed to produce the following pattern? Fill in the spaces with the correct letters to reveal the answer.

2. Show why it is not possible to produce heat from potential energy.

3. What can a bar magnet produce? Use the letters in the box to spell the answer. You must use all the letters.

N	D	I	M
C		L	E
G	T	A	
	F	I	E

4. Devise a way to get ENERGY from the boxlike design.

N	R
G	Y

5. Magnetism is a form of energy. Show how *four* letters of the alphabet are attracted to a magnet.

Section 6

HEAT AND ENERGY

6–1. What Does the Sketch Represent?

Give students an illustration to examine. Ask them to take a light-hearted approach as they record what they think the sketch represents on the line under the drawing. Tell them there is no right or wrong answer.

1.

75

2.

3.

4.

5.

6.

THERMOMETER (stylized as a falling thermometer)

7.

_____

8.

Physical Science

HEAT AND ENERGY (continued)

6–2 Can You Solve the Problem?

The following miniproblems make excellent starters for lessons on HEAT.

1. Mrs. Kelvin, science teacher, enjoyed making up problems and riddles to spring on her classes. She told her classes that temperature can be measured in inches. How do you think she could demonstrate this to her students?
2. Mrs. Kelvin never stops! She said temperature can also be measured in yards. Can you think of a way to do this?
3. Think of a way to cool hot metal with HEAT.
4. Show how the Fahrenheit (F) scale produces 40 percent more heat than the Celsius (C) scale.
5. Give *three* examples of specific heat using one line, two arrows, and one circle.
6. What does the sum of the scrambled letters represent?

   Two *i*'s, three *e*'s, two *n*'s, one *k*, one *t*, one *c*, one *g*, one *r*, and one *y*.

6–3 What's in a Name?

The following activities offer students practice in producing names of people, places, and things from the letters of heat and energy terms.

1. Use the letters in HEAT to write the names of *two* different weather terms for *each* letter. Use H for the first letter in the first term, E for the last letter in the second term, A for the third letter in the third term, and T for the second letter in the fourth term.
2. Use the letters in CONVECTION to write the names of *eight* words that begin with the letter N. You may use a letter more than once.
3. Use the letters in TEMPERATURE SCALE to write the names of five animals with three letters in their names. You may use a letter more than once.
4. Use the letters in the name FAHRENHEIT to write a complete sentence that makes sense. You may use a letter only once. *Exception:* The letter H may be used twice.
5. List three heat and energy terms that name something related to human behavior. The name must be part of the term.

   Example: <u>heat</u> - eat

HEAT AND ENERGY (continued)

6–4 Create-A-Comment

Ask students to examine the two sketches. Then have them write a humorous comment or statement in the bubble next to the figures. Encourage students to write something that relates to the section theme.

1.

2.

HEAT AND ENERGY (continued)

6-5 Riddle Bits

Whenever you wish to change pace or point students in a creative direction, offer them riddles to solve. You can offer an added challenge by setting time limits for responses.

1. If WATER lost three-fourths of its HEAT, what would be left of the WATER?
2. What three-letter combination in HEAT produces HEAT?
3. John did a weird thing. Use the letters in HEAT to reveal his bizarre behavior. You may use the same letter more than once.
4. What kind of energy is needed to develop an idea?
5. How can the first part of an endothermic reaction also be the terminal part?
6. What is the difference between convection and conduction?
7. How can burning be prevented?
8. What can be said about a match that never burns?
9. What is in the middle of an insulator?
10. Why is combustion often called rapid oxidation?
11. Why is it *always* possible to get *thermal* energy from a chemical reaction?
12. Why do sardines make poor thermometers?
13. What part of the human body stays warm?
14. Why are temperatures always hot?

6-6 What Goes Where?

Ask students to solve these puzzles by interpreting symbols, letters, and word hints. Then have them arrange everything in its proper order to reveal the term.

1. Two words

    First word: a + name for sun + flute (–fl) + b.

    Second word: a symbol (0).

2. sounds like the opposite of *out* + Sounds like the word meaning light-colored + Sounds like "site."
3. sounds like C + 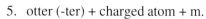 + us.
4. d + a + sounds like "shun" + chemical symbol for radium +
5. otter (-ter) + charged atom + m.
6. "_ _ _ for two" + mug (minus ug) + s.

Physical Science

HEAT AND ENERGY (continued)

6.7 Creative Potpourri

Here are five miniactivities for students to try in the last few minutes of class.

1. There are 15 letters scattered about. Find the 10 letters that, when placed in their correct order, will answer this question: What do you call the transfer of heat energy by movement of liquid?

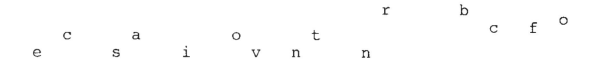

2. There are 20 letters scattered about. Find the 9 letters that, when placed in their correct order, will answer this question: What term describes the transfer of energy waves through space?

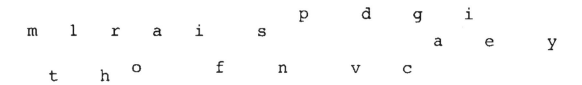

3. Use the letters and numbers in the box to write three terms and two number combinations related to heat. You must use every letter and number in the box.

o	h	e	2	1
m	2	0	r	t
t	h	e	a	l
0	1	t	1	m

4. How many times can HEAT be written from the single- or double-letter combinations located in the box?

H	A
ET	TA

5. Show, using a diagram, molecules on the move.

Section 7

FORCE AND MOTION

7–1 What Does the Sketch Represent?

Let students examine one, two, or three or more sketches. Have them write on the line below each illustration what they think it represents. Ask them to relate their answers to the theme of the section.

1.

2.

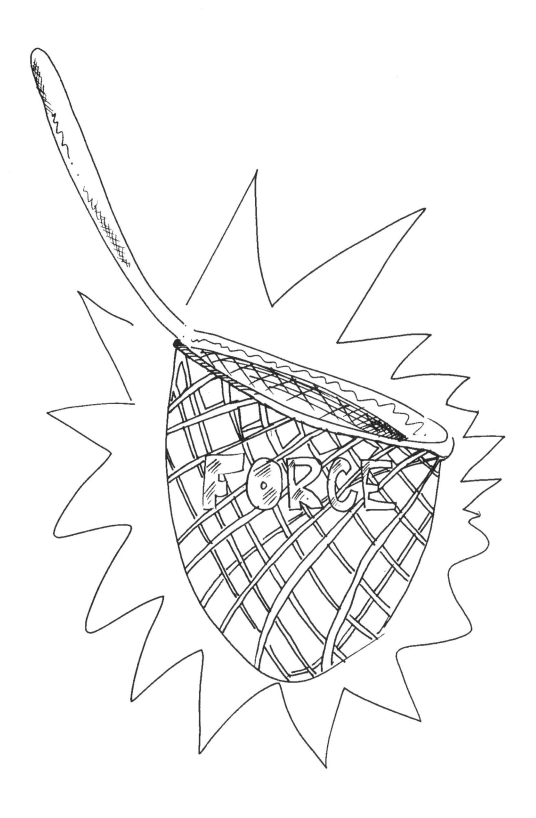

3.

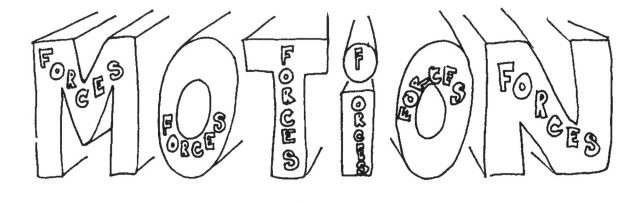

4.

5.

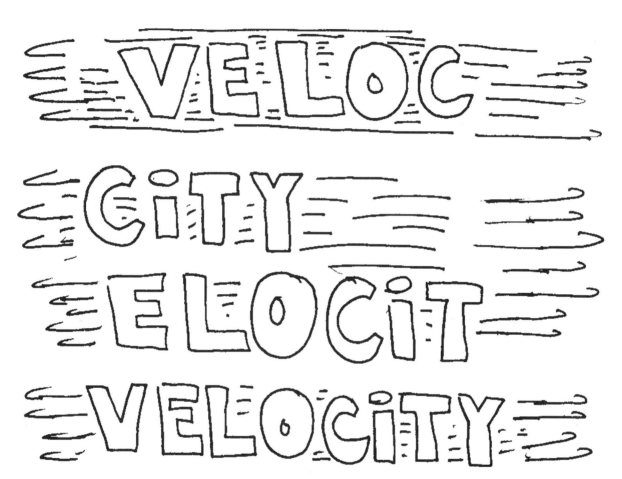

6.

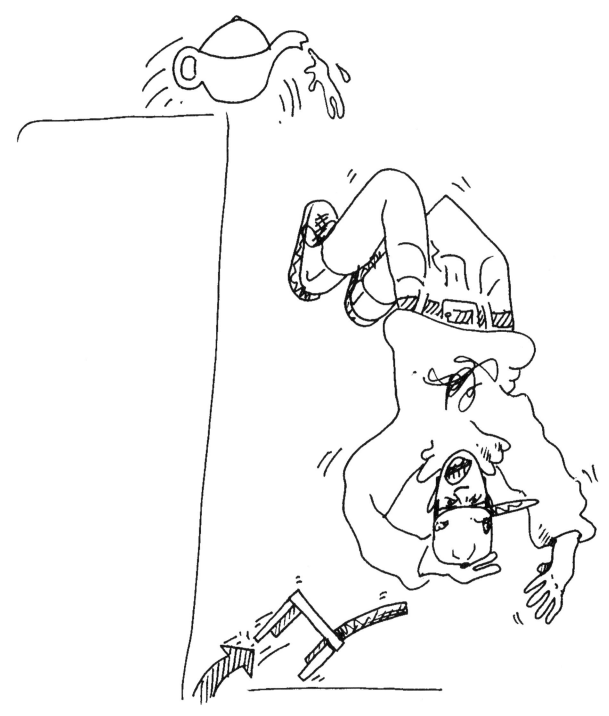

7.

8.

FORCE AND MOTION (continued)

7–2 Can You Solve the Problem?

These six miniproblems require flexible thinking and, at the same time, allow the student to play with a word or two.

1. Show, using two dashes and two arrows, how *equilibrium* might occur with the word EQUILIBRIUM.
2. Show, using FORCES, how forces might produce *friction*.
3. A fulcrum is the pivotal point on which a lever turns. Show, using a sketch, the pivoting point of a fulcrum.
4. Use the letters in the box to produce "a very good formula" for work. You may use a letter only once.

a	i	t
r	e	n
t	c	o
s	e	d
e	i	c
f	m	s

5. Create the name of the man responsible for discovering the theory of gravitation from the items in the box. You may use an item only once.

E	O	B	W	O
L	2	S	O	N

6. Show, using a circle, how MOMENTUM may be a sign of good or evil to come.

7–3 What's in a Name?

Students will have fun creating names, "definitions," and strange sentences in the following activities.

1. List six terms related to force and motion that have a "ti" combination in their names.
2. Create first and last names for each of the listed terms. When read aloud the names should sound like the term.

Physical Science

FORCE AND MOTION (continued)

Example: friction - Rick Shun

velocity   kinetic
inertia   mechanical

3. Make up a boy *and* girl name using the letters of *Law of Conservation of Energy* as the first letters in each name. The first one is done for you.

   L - Lou, Lois   O -
   A -   F -
   W -
                           E -
   O -   N -
   F -   E -
                           R -
   C -   G -
   O -   Y -
   N -
   S -
   E -
   R -
   V -
   A -
   T -
   I -
   O -
   N -

4. Make up sentences using the listed words as part of the sentence. The first one is done for you.

   Example: force

   Sandra wanted to ask *for ce*lery soup for lunch.

   momentum   Newton
   thrust   heat

5. Make up "creative" definitions or examples for each of the listed terms. The first one is done for you.

   Example: Net force - the total effort needed to lift a net

   rate   mass
   weight   speed

FORCE AND MOTION (continued)

7–4 Create-A-Comment

Here are two sketches students can examine. Then, as a test of creative power, have them come up with humorous comments or statements to write in the bubble next to the figures in the sketch. The statements, of course, should match the section theme.

1.

2.

FORCE AND MOTION (continued)

7–5 Riddle Bits

If students solve 6 of the 12 riddles, give them a special award for patience and perseverance.

1. What does VACUM represent?
2. What is the fastest-moving letter of the alphabet?
3. What number *always* has a force acting on it in a gravitational field?
4. How do marching soldiers travel?
5. What one thing is always true about a force?
6. What is "equal" about equilibrium?
7. Why is it hard for a fulcrum to be involved with balance?
8. When do parents use the "Law of Moments"?
9. What does this mean: ғʀɪCTION?
10. How are *speed* and *acceleration* related?
11. What does this mean: ecrof?
12. What force and motion term is related to a rodent?

7–6 What Goes Where?

Have students decipher symbols and phrases. Then ask them to unscramble everything to identify the mystery terms.

1. en + rhymes with gum (three-letter word) + not dad, but ___.
2. 1 *cent* + 1 c + 1 (third letter of alphabet) + 1 sounds like "see."
3. "____ and behold" + San Francisco, for example, + ev (backward).
4. 23rd letter of the alphabet + sounds like mate or fate.
5. Two words

   First word: + or –.

   Second word: Sounds like nation + elecc (backward) + a.
6. Two words

   First word: ↻, ↻, ↻

   Second word: Sounds like lotion.

Physical Science

FORCE AND MOTION (continued)

7–7 Creative Potpourri

Here are five miniproblems for students who wish to practice flexible thinking.

1. You have three blocks. They are $\boxed{R}$ $\boxed{B}$ $\boxed{U}$. Show, using arrows, how you can arrange the blocks to produce friction. Also, provide a brief explanation for your strategy.
2. Show how *six* elements properly arranged can produce *friction*.
3. Give an example of an *unbalanced force* using a five-lettered word.
4. Show how *air resistance* can be eliminated using friction.
5. Show how horizontal velocity can be changed to vertical velocity.

Section 8

MACHINES

8–1 What Does the Sketch Represent?

Students may tackle one, two, or all eight illustrations. Regardless, have them examine each sketch and write what they think the drawing represents on the line below the sketch.

1.

2.

3.

4.

5.

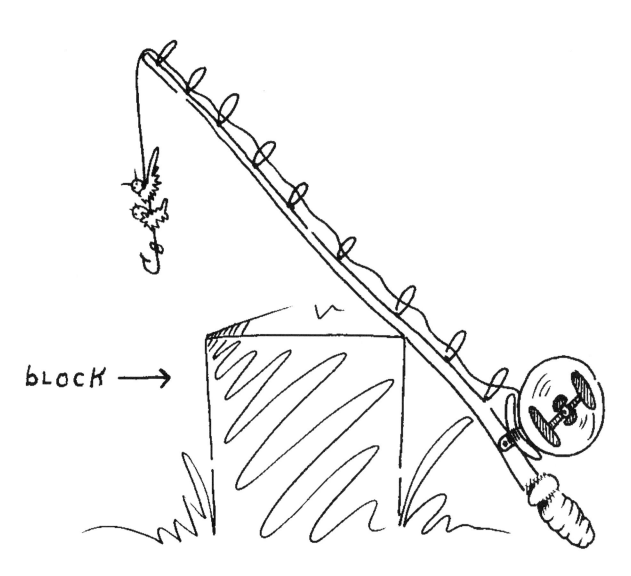

6.

7.

_____

8.

MACHINES (continued)

8–2 Can You Solve the Problem?

Here are five miniproblems designed to keep a student's think tank moving.

1. Some say MACHINES make work easier; others say there is nothing *easy* about a machine. Well, there isn't! Think of a way to show this.
2. There are several examples of simple machines. Use four of the letters in *simple* to serve as the first letters in four examples of simple machines.
3. Show an example of work input using the words WORK INPUT.
4. Machines that combine two or more simple machines are called compound machines. Create a compound machine out of the word *incline*.
5. Mechanical advantage or MA is the number that shows how many times a machine multiplies the force applied to it. Indicate with a line, arrows, or a circle where the MA is located on all MACHINES.

8–3 What's in a Name?

These five activities allow students to find names hidden in terms related to machines.

1. A machine, like a lawn mower, can produce a continuous dull pain if used for a long period of time. Find the letters in MACHINE that reveal the name of the culprit responsible for this condition.
2. Use only three letters in the word LEVER to write a three-lettered boy *and* girl name.
3. Write the name of an animal that rhymes with each of the terms related to machines. The first one is done for you.

   Example: screw - shrew

force	torque
wheel	plane
bar	incline

4. Use the letters in the box to spell the names of three simple machines. All letters must be used.

   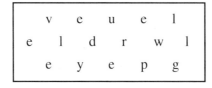

5. Use the letters in FRICTION to spell the names of *five* items that can produce or be created by *friction*. You may use a letter more than once.

Physical Science   111

MACHINES (continued)

8–4 Create-A-Comment

Let students examine the two illustrations. Then ask them to write a humorous statement or comment in the bubble next to the figure. The statement should relate to the theme of the section.

1.

2.

Physical Science

MACHINES (continued)

8–5 Riddle Bits

Here are 12 riddles for students to groan over. They do wake up the neural center, especially during the first few minutes of class.

1. What lies next to the edge in a wedge?
2. What kind of machine can never have children?
3. What does an older lever have that a younger lever doesn't?
4. Gene Machine didn't like his name. Kids made fun of it. When he grew up he changed his last name and nobody poked fun at him again. What do you think was his new last name? *Hint:* No letters in his original name were changed.
5. Why did the scissors refuse to fly coach?
6. What are the two parts of a lever?
7. What college degree does a lever have?
8. What kind of machine weighs 16 ounces?
9. What part of a *force* is missing in an *effort?*
10. What kind of a machine is a *whaxle?*
11. They say that half of all machines are made up of a front part of the lower jaw. How could this be?
12. How many ants, 10 cm in length, can you get on a 30 cm long slanted ramp?

8–6 What Goes Where?

Here are six activities for students who like to unscramble symbols, words, and numbers to identify correct terms.

1. 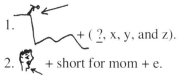 + (<u>?</u>, x, y, and z).
2. 🧑 + short for mom + e.
3. Adam and _ _ _ + r + l.
4. 🐟 + e + sounds like "Ian" + pronounced "see."
5. Two words

    First word: 👣 + w

    Second word: 1 + e + 🔨 (rhymes with *stacks*).
6. Rhymes with brick + sounds like "shun."

MACHINES (continued)

8–7 Creative Potpourri

Here are five activities that provide students with a chance to give the "challenge center" in their brains a nudge or two.

1. What two compound machines crossed to produce the following pattern? Fill in the spaces with the correct letters to reveal the answer.

2. Show, using symbols and letters, what kinds of lever an athletic referee might be.
3. Pete Punter, kicker on the school football team, is called "a first-class player with a third-class leg" by his teammates. Show, using symbols and letters, what his teammates mean by this.
4. A level is a horizontal line or surface. Show, using letters and arrows, how a level can be warped or bent and still be level.
5. There are six terms related to machines. Write a rhyming descriptive word for each term.

   Example: wheel peal

   force   fork
   device   pulley
   spoon   weight

EARTH SCIENCE

- ▼ ENERGY SOURCES
- ▼ ROCKS AND MINERALS
- ▼ VOLCANOES AND EARTHQUAKES
- ▼ FOSSILS AND GEOLOGIC TIME SCALE
- ▼ THE EARTH'S FORCES
- ▼ WEATHER AND CLIMATE
- ▼ ASTRONOMY
- ▼ OCEANOGRAPHY

Section 9

ENERGY SOURCES

9–1 What Does the Sketch Represent?

These six drawings provide a fun, relaxing way for students to think about a subject. So have them examine each sketch and write what they think the drawing represents on the line below the sketch.

1.

2.

3.

4.

5.

6.

Earth Science 123

ENERGY SOURCES (continued)

9–2 Can You Solve the Problem?

The following five activities give a student's brain power a chance to expand. The effort will require a modicum of energy.

1. If an energetic person lost his or her "energe," what might develop?
2. The four stages of coal—peat, lignite, bituminous, and anthracite—fit into four slots in the puzzle. Find the correct slots. Then answer the question: What seems to be wrong with the puzzle?

3. What is the energy source able to move generators and produce electricity? The answer is hidden in the four symbols.

4. Darken the letters of three terms related to natural energy sources in the puzzle. The terms may be forward, backward, or up and down. Then answer the question: After you shade in the three terms, what will you have left?

T	L	L
A	I	A
E	O	O
P	O	C

ENERGY SOURCES (continued)

5. Use all the letters scattered inside of the circle to spell OIL. How many times can you spell OIL?

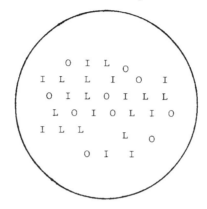

9–3 What's in a Name?

Students will have their hands full locating terms to match names of common objects. In most cases, an active imagination is the key to success. Also, they will be busy looking for the right term or word to complete an activity.

1. List *four* names of animals from the letters in GEOTHERMAL. *Rule:* You may use a letter no more than two times.
2. Find *two* names of animals from the letters that *do not* form the word GEOTHERMAL. *Rule:* You may use a letter more than once.
3. Find a term related to energy that rhymes with each of these 15 listed words.

toil	mule	bass
linoleum	fleet	illusion
dynamite	mole	decision
luminous	polar	sunflower
friend	seam	daughter

4. Fossil fuel is a mineral fuel found in the Earth's crust. Coal, a fossil fuel, is a combustible mineral resulting from prehistoric vegetation becoming squeezed, thus changing into a solid burnable substance.

   Write the name of an object, event, or phenomenon related in some way to the coal formation process next to the letters in the word FOSSIL. The names chosen must begin with the letters that spell FOSSIL.

EARTH SCIENCE

ENERGY SOURCES (continued)

F _____
O _____
S _____   FUEL
S _____
I _____
L _____

5. A hydrocarbon is a substance made up only of carbon and hydrogen. Gasoline, for example, comes from petroleum. Petroleum is a mixture of hydrocarbons.

   Two of the five hydrocarbon formulas listed are hidden in the puzzle. When you find them, circle the names of the formulas on the list. You must use every letter in the puzzle.

H	H	H	H
H	C	H	C
H	H	■	H
C	H	H	C
H	C	H	H

Hydrocarbon	*Formula*
methane	CH_4
ethane	C_2H_6
butane	C_4H_{10}
pentane	C_5H_{12}
hexane	C_6H_{14}

   Now answer these questions:
   a. If the black square was an H, how would it change the answer?
   b. If the black square was a C, how would it change the answer?

6. Write the names of three fossil fuels from the letters scattered in the diagram of a Glossopteris, a seed-fern with thick bladelike leaves.

ENERGY SOURCES (continued)

9-4 Create-A-Comment

Students need to conserve their energy and warm up their brain cells before taking on this section. Once they prepare themselves mentally, give them the two drawings. Have students examine each one and write a comment or statement in the balloon above the figures to match the theme of the illustration.

1.

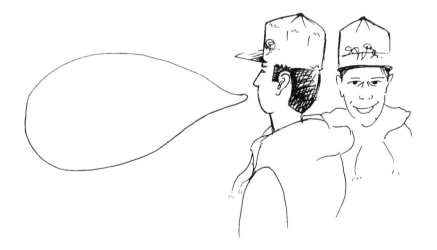

2.

ENERGY SOURCES (continued)

9–5 Riddle Bits

Prime your students for a round or two of riddles. Solving riddles hones creative thinking strategies.

1. What do swamps and amperes have in common?
2. What stage of coal development occurs at darkness?
3. Why did some plants turn into coal?
4. What do the fourth, fifth, sixth, and seventh letters play in the word PETROLEUM?
5. Where do hydrocarbon molecules swim?
6. Some people have trouble understanding the difference between fusion and fission. If you wanted to keep things in turmoil, what could you do?
7. Scientists created a synfuel (synthetic fuel) and called it PROCON. Twelve thousand people reported on the quality of the product. What do you think were the results?
8. *Tea* makes up 60 percent of a power source. What is the power source?
9. Why is wind power considered important for our future energy supply?
10. Why can't you get oil from THE SKY?
11. What do you call petroleum without manners?
12. What kind of energy is needed by a person whose job it is to greet people all day?

9–6 What Goes Where?

Have students unscramble the symbols and words to identify the correct terms for the following seven items.

1. opposite of out + d + w.
2. eat (past tense) + r + w.
3. d + s + another word for sick + opposite of lose + m.
4. (second word)   (first word)
    + s + sp   opposite of cold (*two* words).
5. (second word)   (first word)
        r + Roman god of sun + a (*two* words).
6. (first word)   (second word)
   s + prefix for life + opposite of pa + s   l + sounds like "few" + s (*two* words).

Earth Science

ENERGY SOURCES (continued)

7. (first word)
    w + p + er + o   (second word)   (*two* words).

9.7 Creative Potpourri

The six items in this section allow students to play with words, examine a pun, and just have some low-key, relaxing fun.

1. A pun plays upon words similar in sound, but having a different meaning. Read the following pun statement and record what you think it means.

    Sign reads: For peat's sake, help conserve energy.

2. Use the words in the list to create fictitious sporting or recreational events.

    Example: Coal Bowl - Miners play in a football game.

    | oil | fission |
    | fuel | solar |
    | wind | peat |

3. How about a challenge? You have one minute to change the first letter in *two* of the terms and discover *two* water-dwelling organisms.

    | wind | lignite |
    | bituminous | carbon |
    | nuclear | shale |
    | tar | sediment |
    | fuel | water |

129

ENERGY SOURCES (continued)

4. Darken the letters that spell COAL six times. If you do this correctly, the unshaded letters will spell the name of a fuel product from the Pennsylvanian period.

C	C	A	L	A	C	T	O	C
L	O	A	L	S	M	A	L	A
O	E	O	I	O	C	O	L	I
O	A	L	C	L	S	A	C	X

5. Show, with a diagram, how to connect the four stages of coal development in order from left to right.

6. Several years ago a slogan appeared encouraging people to conserve energy. It read:

    "Don't be fuelish. Conserve energy."

Make up two or three slogans urging a commonsense approach to using fossil fuels.

Section 10

ROCKS AND MINERALS

10–1 What Does the Sketch Represent?

The eight drawings are designed to challenge the creative "thinking neurons" of students. Each sketch should elicit a fun, albeit slightly bizarre, response.

Ask students to examine each sketch and have them write what they think the sketch represents on the line below the illustration.

1.

131

2.

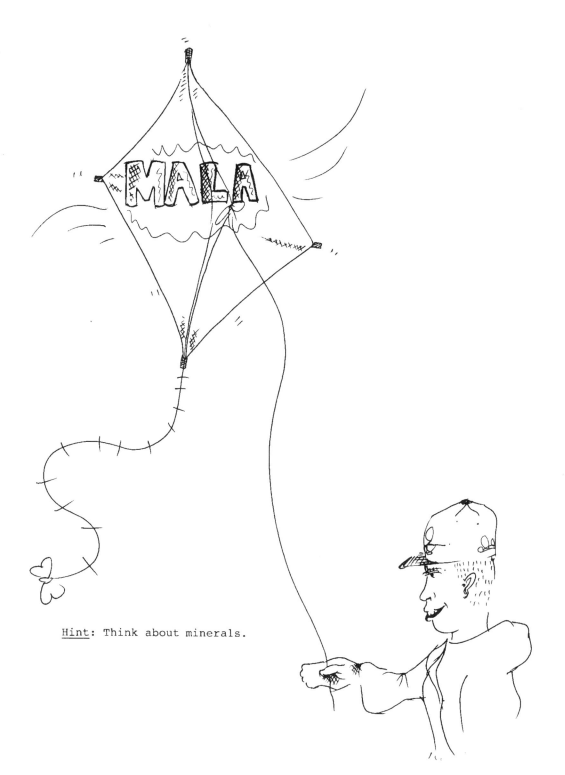

Hint: Think about minerals.

3.

molehill

4.

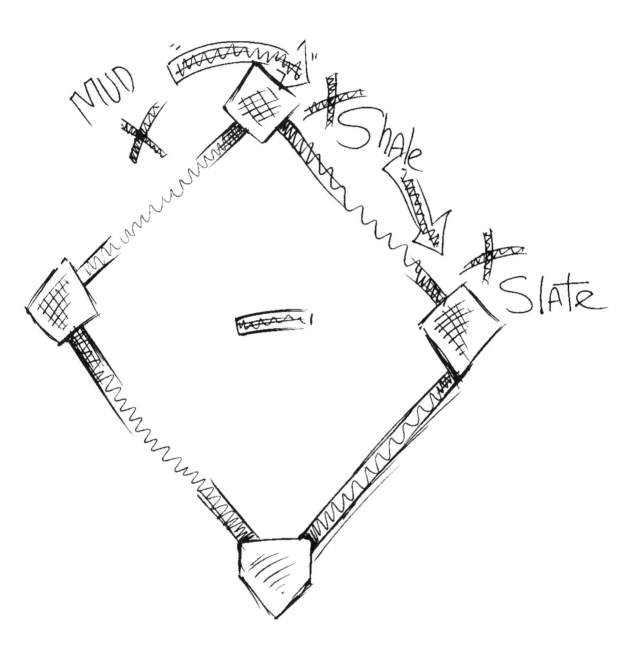

5.

Hint: Largest group of minerals.

6.

Hint: Think about minerals.

7.

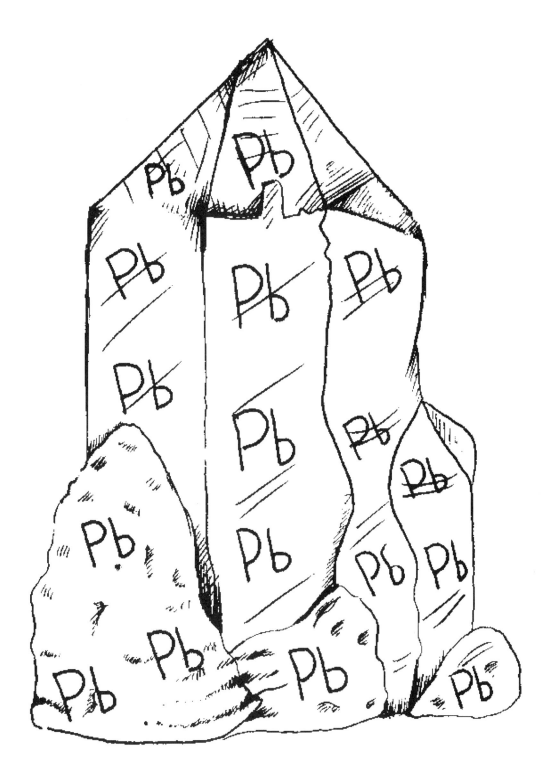

8.

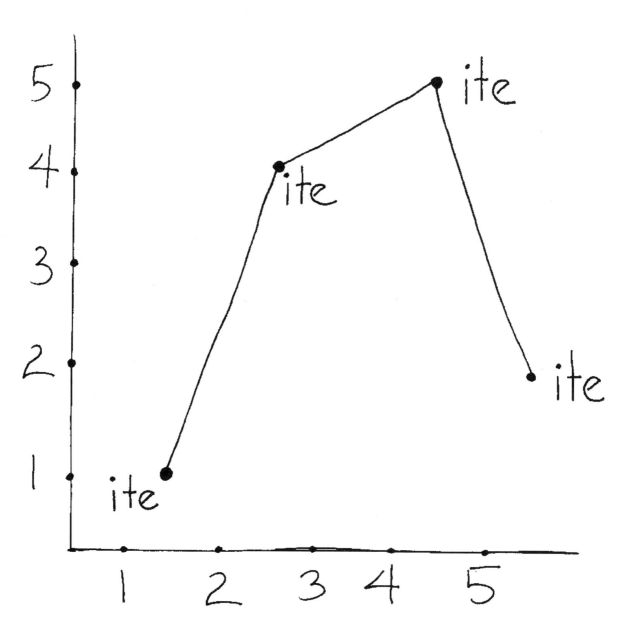

Earth Science

ROCKS AND MINERALS (continued)

10–2 Can You Solve the Problem?

Turn students loose on the six miniproblems. If they rev up their thinking machines, they should enjoy the trip. Admittedly, these are tough items and require both patience and determination.

1. Think of a way to make GOLD out of LEAD.
2. Mr. Sodium was a nice person but really "backward." He married his girlfriend, Chlorine. They had a child. What was this child considered? *Hint:* Think of the formula for table salt.
3. Show how you can complete the word ROCK by using potassium, carbon and oxygen.

    R __ __ __

4. Show how a rock might look under tremendous pressure.
5. Show what part of shale, original sedimentary rock, undergoes the most pressure before turning into slate, metamorphic rock.
6. The list contains eight rocks and minerals with missing letters. Four of them have two missing i's. Put the double missing i's in their correct places. Leave the rest of the names as they appear.

    a. F__ L S__TE
    b. D__A B__SE
    c. D__OR__TE
    d. G__P S__M

    e. PER__DOT__TE
    f. M_CROCL__NE
    g. PYR__X__NE
    h. L__MON__TE

10–3 What's in a Name?

Students will stay sharp by matching terms to rhyming names and tossing around a few nouns or verbs. These brain-stroking activities keep the neurons charging ahead.

1. See how many nouns you can list out of the rock and mineral terms. You cannot move letters around, but you can use a letter more than once.

    Example: quartz - quart or art

    feldspar   sandstone
    hematite   conglomerate
    serpentine   quartzite
    sedimentary   dolomite
    anticline   syenite

2. Write the name of a rock or mineral that has the following listed groups of letters in its name.

ROCKS AND MINERALS (continued)

Minerals	Rocks
tz	tz
lc	bb
yp	yr
ph	cc
nn	ss

3. Name an animal that relates in some way to each of the listed minerals. *Briefly* describe the relationship.

Mineral	Animal	Description of Relationship
wolframite		
chromite		
sulfur		
realgar		
serpentine		
stilbite		

4. Try to find action words hidden in the listed terms. Do not move letters around to form words.

   Example: hornblende - blend.

brown coal	pegmatite
cassiterite	plagioclase
clay	travertine
feldspar	uranium
corundum	

5. Match the mineral name in the left-hand column with the "made up" description in the right-hand column. Connect the name and description with a line.

Name	Description
a. agate	A. daughter of your brother or sister
b. realgar	B. the saline content of a recess in the shore of a sea.
c. selenite	C. a large entrance or passage.
d. pumice	D. a genuine snakelike fish with sharp teeth
e. basalt	E. small rodents known to infest the city of Pu.
f. gneiss	F. a person who sells merchandise after dark.

Earth Science

ROCKS AND MINERALS (continued)

6. Write the names of eight minerals that make up rocks on the lines provided in the diagram. The mineral names *must* begin with the letter located in the box touching the lines. Two examples are provided. There are three lines touching R, one line touching O and C, and three lines touching K.

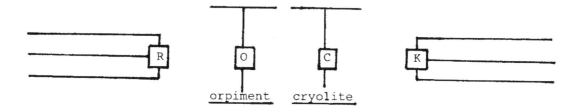

7. Write the names of nine minerals on the lines provided in the diagram. The mineral names *must* begin with the letter located in the box touching the lines. Five of the mineral names appear on the diagram.

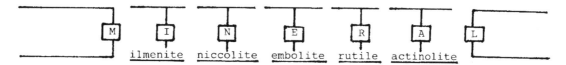

8. Letters that make up *four* igneous rock names ending in "ite" are scattered inside of a rock diagram. Use every letter in the diagram to form the names of the four rocks. Place answers in the spaces below the diagram.

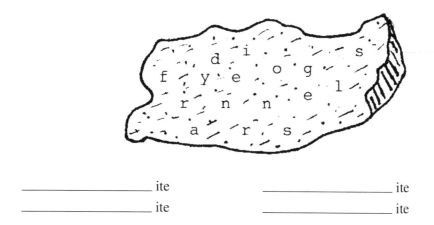

_____ ite   _____ ite
_____ ite   _____ ite

9. Write the name of the rock in the empty space that rhymes with the two words to the left and to the right of the space.

ROCKS AND MINERALS (continued)

nail	_____	pail	garble	_____	warble
rate	_____	gate	rough	_____	stuff
list	_____	mist	talk	_____	stock
peace	_____	lease	flirt	_____	skirt

Earth Science

ROCKS AND MINERALS (continued)

10–4 Create-A-Comment

Once students gird themselves to think about rocks and minerals, give them the following two drawings to study. As they examine each copy, have them write a comment or statement in the balloon above the figures to match the theme of the illustration.

1.

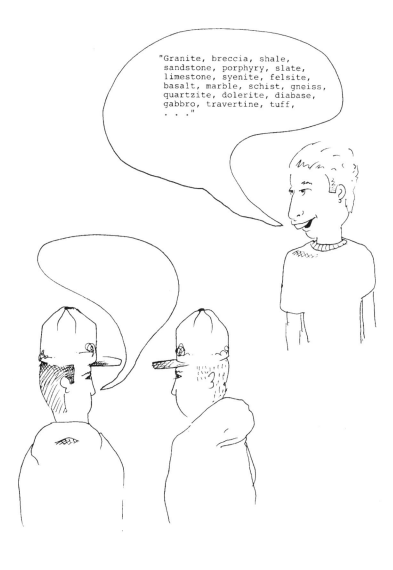

2.

Earth Science

ROCKS AND MINERALS (continued)

10–5 Riddle Bits

It's time for more riddles. So tell students to relax, have fun, and take a chance on the wild side.

1. If you were born on April 1, what would be your birthstone?
2. Where would a geologist find herself if she caught her foot under a boulder?
3. What mineral is not to be trusted?
4. What mineral is never wrong?
5. What mineral is related to eating patterns?
6. What mineral is the most esthetically pleasing?
7. What mineral is known as "the boxer of the mineral kingdom"?
8. Mr. Crystal, geology teacher, was heard babbling these words out loud: "There are three unequal axes, two of which are at right angles while the third is inclined to the vertical." They came and took him away. Where did they take him?
9. Mr. Bonkers, a weird science teacher, wanted students to identify minerals by his behavior in class. As an example, he held up a greenish-red crystal and said, "This crystal has a hardness of 7 to 7.5. If it were a city by the name of Maline, we could go sight-seeing." What was the mineral?
10. What mineral was named after Popeye's girlfriend?
11. Mr. and Mrs. Morphic move next door to you. You see Mr. Morphic unpacking his car. You go up to him and introduce yourself. What have you done?
12. A cement contractor poured a new driveway for Mrs. Dawson. A mysterious event happened: Mrs. Dawson counted 23 "ion" markings in the fresh cement the next morning. What do you think they are?

10–6 What Goes Where?

Let students wrestle with a mixture of symbols, words, and numbers. Have them unscramble the mesh and locate the correct terms for the seven items.

1. ene + 🪨🪨 + π. *Hint:* A mineral.
2. 〰️ + us + ig + o.
3. opposite of wrong + π. *Hint:* A mineral.
4. s + opposite of short + i's + 👁️👁️
5. s + 〰️ + s *Hint:* A mineral.
6. e + 2000 pounds + s.
7. e + chemical symbol (Sb) + 🐍

ROCKS AND MINERALS (continued)

10–7 Creative Potpourri

Three words describe this section: strange and more strange. Give students freedom to go on a creative binge. Let them design their own humor to match each item.

1. Use each word as the key meaning to an imaginary conversation. Here's an example of one person speaking to another:

    Example: "Albite (Al - I'll, bite - bet), you can't jump over the fence, Cheryl."

granite	porphyry
gneiss	chert
breccia	chalcedony

2. Make up *two-word* descriptions for each of the listed items. Remember, the zanier the better.

    Example: selenite - an "ite" salesperson

marble	agate
stalagmites	sediments

3. List *two* expressions that use rock or rocks to describe their hardness.

4. Make up two or three car bumper stickers that relate to rocks and minerals.

    Example: METAMORPHISM IS UPLIFTING.

5. A pun plays upon words similar in sound, but having a different meaning. Read the following pun statement and record what you think it means.

    Have you heard the new rock group: Mica Schist and the Foliations?

Section 11

VOLCANOES AND EARTHQUAKES

11–1 What Does the Sketch Represent?

The following seven sketches will allow students to respond in an open-ended manner and not be concerned with a right or wrong answer. Any legitimate response will guarantee success.

Have students study the sketch and write what they think the sketch represents on the line below the drawing.

1.

2.

3.

4.

5.

6.

CRUST

_____

7.

VOLCANOES AND EARTHQUAKES (continued)

11–2 Can You Solve the Problem?

These problem items will wake up the neurons and provide a comfortable challenge for creative thinkers.

1. Parts of two terms are mixed together. Separate the two terms from each of the following words:

   lagma   truption   epicus   badrock

2. Make a list of eight words that express how people react when they experience an earthquake.
3. Five elements make up the majority of basalt, most common volcanic rock. Three of them are iron (Fe), aluminum (Al), and magnesium (Mg). The remaining two elements are hidden in the following sentence. See if you can locate the two elements. Write each element's name and chemical symbol in the spaces provided.

   Sidney ran into the room shouting, "Carla, guess what? I've got some great news for you."

   Name of Element   *Symbol*

   _____   _____

   _____   _____

4. There are enough letters scattered inside of the sketch to spell MAGMA *five* times. See if you can find all of them.

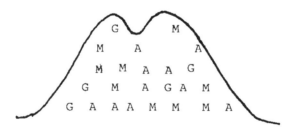

5. The letters that spell the names of three different kinds of lava are scattered along the sides of the volcano. Unscramble the letters and identify the names of each lava. *Clues:* One word rhymes with meridian, one contains the plural of mouse, and one sounds like "end in sight."

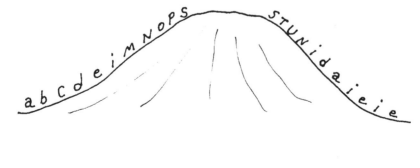

Earth Science

VOLCANOES AND EARTHQUAKES (continued)

6. These figures illustrate an explosive phenomenon known to people on Earth. What is it?

11–3 What's in a Name?

Students can immerse themselves in terms and names in this section. Once again, creative energy is set free to keep the cerebral wheels turning.

1. The zone between the crust and mantle is known as the asthenosphere. There is a name of an animal and a boys' name hidden in the term. Can you find them?

2. Try to name the eight volcanoes by using words that sound like their names and the location of each volcano. Write the name of the volcanoes in the spaces provided.

Sound-Alike Words	Location	Volcano
crack a toe	Indonesia Archipelago	_____
count paint melon	Washington (USA)	_____
kilam and jaro	Kenya, Africa	_____
pair of cooteen	Mexico	_____
strong bowlee	Italy	_____
kill a waya	Hawaii	_____
cert see	Iceland	_____
vee voo veyas	Italy	_____

3. An earthquake occurs when underground rocks break. The point of breakage is called the *focus*. List *six* two-word combinations to describe the focus. Use rock as one of the two words. Be creative.

    Example: rock rupture

4. Create a rhyming word description for each of the following definitions. The first one is done for you. (This is a toughie.)

    Example: A person who dislikes volcanoes - crater hater

VOLCANOES AND EARTHQUAKES (continued)

a. An earthquake with an unknown origin.
b. Black, fine lava spread over a large area.
c. The ground swaying back and forth.
d. A person whose apartment lies over an active fault line.

5. Richter and Mohorovicic are scientists who have made significant contributions to seismology. Use the letters in their names to spell *three* terms related to seismology. You may use a letter more than once.

Earth Science

VOLCANOES AND EARTHQUAKES (continued)

11–4 Create-A-Comment

Student imaginations tend to run wild when they make up their own comments or statements to match a sketch's theme. This is an effective way to make an activity fun and keep it moving in a smooth, relaxing manner. So, once again, give students a copy of a drawing and let them create a humorous comment or statement to place in the bubble above the figures in the sketch.

1.

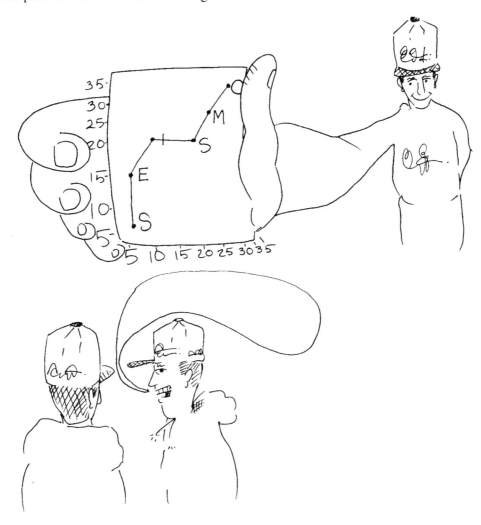

2.

Earth Science

VOLCANOES AND EARTHQUAKES (continued)

11–5 Riddle Bits

Let your students tackle a few riddles. Riddles are wonderful stimuli designed to stretch the thinking fibers to their limits.

1. What German scientist coined the expression "Get my drift?"
2. What would an earthquake be if it occurred in the same location several times?
3. What happens when the Earth's stable crust begins to rattle and shake?
4. What is a geologist's favorite dinnerware?
5. What insect is trapped in the layer of the Earth located next to the crust?
6. What do you call people who dislike volcanoes?
7. What is a seismologist's greatest fear?
8. What is a seismologist's favorite social drink?
9. What would be an adequate two-word description for the focus of an earthquake?
10. What would be an adequate two-word description for the odiferous gases given off by an erupting volcano?
11. What is a volcanologist's favorite dessert?
12. What does a golfer who misses hitting the ball off the tee and a person who feels minor tremors before an earthquake have in common?
13. Where is the coldest part of an epicenter?
14. A "friendly" volcano erupted in Mexico. It didn't cause any bodily harm or property damage. What did the people name the volcano?
15. What do you call an extinct volcano?

11–6 What Goes Where?

Have students unscramble and decipher the words and phrases. Then, as an added challenge, have them combine the words and phrases with the numbers and symbols to reveal the correct terms.

1. c + Sb (chemical symbol) + ex + t.
2. m + ic + 👁 '+ s.
3. am + ☀ + t + i.
4. s + 〰️ + ol + v.
5. r + r + 4 plus 4 + c.
6. t + 🏔️

VOLCANOES AND EARTHQUAKES (continued)

7. c + k + 4 + sh + o.
8. ~~~ + 👂 + sh (two words).
9. ry + 👁 + opposite of pa + pr + ~~~ (two words).
10. s + m + 🌳 + or.

11-7 Creative Potpourri

This section offers students a collection of puns, expressions, hidden words, and a miscellaneous activity to satisfy the creative urge.

1. A pun plays upon words similar in sound, but having a different meaning. Read the following two pun statements and record what you think they mean.
   a. Did you hear about the seismologist who was fired for losing his focus?
   b. And what about the unhappy volcanologist? He just couldn't "go with the flow."
2. Think of four or five expressions that relate to an active volcano.
   Example: Lora really blew her top at Lesley.
3. What three words are hidden in the word SCARP. *Rule:* You can use a letter more than once, but you can't move letters around. Make a sentence using the three words.
4. Naive means to have a simple nature, that is, a lacking of careful analysis. For example, a naive person might think *rhyolite* (volcanic rock) is a type of diet cracker.

See how many naive descriptions you can create from the listed terms. Begin each statement as follows: A naive person thinks . . .

bedrock	crater
epicenter	scoria
Krakatoa	tremor

Section 12

FOSSILS AND GEOLOGIC TIME SCALE

12–1 What Does the Sketch Represent?

Each sketch offers enough clues to help students reveal the meaning of the sketch. Some responses require a student to stretch the neurons to their breaking point. Have students examine each sketch and write what they think the sketch represents on the line below the drawing.

1.

2.

3.

4.

5.

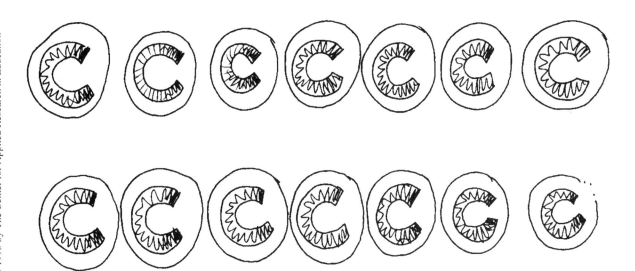

6.

7.

8.

Earth Science

FOSSILS AND GEOLOGIC TIME SCALE (continued)

12–2 Can You Solve the Problem?

These fun problems should be challenging enough to fire up a neuron or two. And, of course, a student who keeps the thinking process flexible will find success.

1. Use the letters in SEDIMENTATION to write the names of four items found on the human body. A letter may be used more than once.
2. List ten words related to fossils and geologic time scale that have the "at" letter combination in them.
3. Mrs. Martinez, science teacher, has 20 students in her class. She teaches in a small room, five rows of seats, four seats per row. She made a seating chart using the *Principle of Superposition* to assign her students by their ages.

    Sketch the seating chart and place the students on the chart according to age. The students' ages are as follows: 14, 14.6, 14.1, 13.9, 15, 15.2, 15.3, 15.6, 15.1, 13.8, 14, 14.9, 14.7, 15.3, 15.4, 15.8, 16, 15.5, 14.9, and 15.2. Arrange students from left to right on the seating chart.
4. Match the *fossil record* examples with the correct *geologic era* using connecting lines. The first one is done for you.

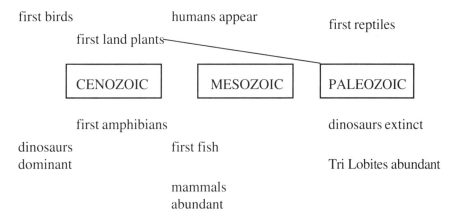

5. Show, by using the letters NGOIRAC how ORGANIC evolution might occur.
6. How could you show a geologic cross section using two words: geologic and section?
7. How could you show an example of Punctuated Equilibria using only one word: equilibria?
8. Hardened tree sap is called amber. Insects over a million years old have been trapped in amber. Write the names of insects that might be found in amber. Insect names *must* begin with the letters that spell AMBER.

    Example: A - ant or aphid
9. Show how many bones (🦴) it would take to make a fossil.

FOSSILS AND GEOLOGIC TIME SCALE (continued)

12-3 What's in a Name?

This section offers practice in finding words that rhyme and matching boy/girl names with scientific terms. Give your students a chance to show how clever they are.

1. A combination of two girl names sound like the term for deterioration or rottenness. What are the two names?
2. What girl's name sounds like half the word for reptiles that flourished during the Mesozoic Era?
3. Use the three clues to identify a girl's first name: insects, tree, and sticky.
4. What boy's name appears in the name of an extinct animal found in the LaBrea Tar Pits of Southern California?
5. What boy's name appears in the name of an epoch during the Cenozoic Era?
6. Find matching boy or girl names/nicknames for the organisms found as fossils.

    Example: Grant Ant.

fig	tick
clam	fish
bee	fly
snail	stem

7. What geologic era describes a "zoic" with a faint luster?
8. What 15 girl's names can you get from the letters that spell PENNSYLVANIAN PERIOD? A letter may be used more than once.

Earth Science

FOSSILS AND GEOLOGIC TIME SCALE (continued)

12-4 Create-A-Comment

It's time to be supercreative again. Give students copies of each drawing and have them write a comment or statement in the balloons above the figures in the illustrations. Students should have little difficulty finding comments to match this popular theme.

1.

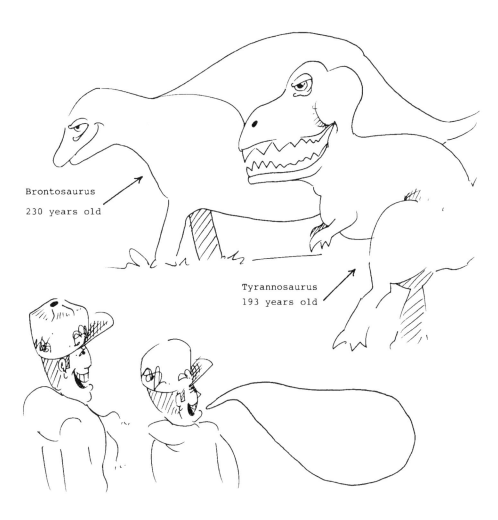

2. DINOSAUR MUSEUM

FOSSILS AND GEOLOGIC TIME SCALE (continued)

12–5 Riddle Bits

Now is a wonderful opportunity to riddle your students down to size. Oh, well, ... at least this section provides a rest stop along the road to academic splendor.

1. What's happening in the sketch?

2. What example is being shown in the sketch?

3. What sea animal has several ears but cannot hear?
4. What does the sketch show?

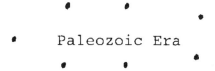

5. What ancient animal of the Ice Age has a name similar to a mother insect?
6. What type of fossil records carry the name of an automobile?
7. Fossil bones were unearthed recently. Nobody knew for sure what animal died and became fossilized. However, the word VERTEBRATE gave the key to its identity. What was the animal?
8. There were two identical twin brothers, both paleontologists, whose last name was Eon. Their parents were poetic people. What were the Eon brothers first names?
9. The Eon brothers had two identical twin girlfriends, both stratigraphers, whose last name was Era. Their parents were poetic individuals. What were the Era sisters first names?
10. Why do some people think paleontologists are weird?
11. How does the word PALEONTOLOGIST represent two or more eras of time?
12. If a newspaper person wrote articles on deposition from the beginning of time, where would these articles appear in the newspaper?

FOSSILS AND GEOLOGIC TIME SCALE (continued)

13. If your dog, Fang, died and became fossilized, in what final condition might somebody find the animal?
14. How many letters would be in the words FOSSIL BONES if you took one away?

12–6 What Goes Where?

Now it's time to unscramble words, numbers, and symbols. Then, of course, if everything goes well, the correct term will appear.

1. nt + (10¢) + se.
2. opposite of off + 5th letter.
3. end of sentence (.) + 4 s' + 4 i's + 2 p's + man (two words).
4. ◇ ◇ + rep.
   (floor covering)
5. 19th letter + California (abbr.) + t.
6. life → lif → li.
7. w + unit of electrical current (abbr.) + s.
8. sodium (chemical symbol) + (¢) (minus the first r) + ry.

12–7 Creative Potpourri

This is what students have been waiting for: a mixture of puns, silly definitions, and a bizarre look at ways to treat an academic subject.

1. A pun plays upon words similar in sound, but having a different meaning. Read the following pun statements and record what you think they mean.

   a. Did you read the book about how some insects become fossilized? It's titled *Forever Amber.*

   b. A weird thing happened. Seventy million years ago an ant, a bug, and a fly—all living in an empty snail shell—were buried and became fossilized. A paleontologist found the specimen and decided to write a play about it with three in the cast.

2. "Fossil Records" has just signed three new rock bands. The first group is called "Emma Knight (ammonite) and the Nautiloids." Make up names for the other two groups.

3. Create silly definitions for each of the terms.
   Example: era - the mixture of gases that make up the Earth's atmosphere

   crinoid   trilobite
   strata   nautilus
   guide fossils   extinction

FOSSILS AND GEOLOGIC TIME SCALE (continued)

4. Make up two or three personalized car license plates relating to fossils or geologic time scale. Here are two examples:

   | FSL MAN | | I DIG FSLS |

5. Create two or three slogans or expressions that relate to fossils or geologic time scale.

   Example: SEDIMENTATION produces good STRATA-GY.

6. Combine two or more rhyming words that relate to fossils or geologic time scale.

   Example: docile fossil or swell shell

Section 13

THE EARTH'S FORCES

13–1 What Does the Sketch Represent?

It's time to give the imagination a wake-up call. So have students examine each sketch and record what they think the sketch represents on the line below the drawing. Once again, alert students to match comments with the section theme. And, of course, a splash of humor helps liven up the activity.

1.

2.

3.

Temperature and Pressure

4.

5.

SOLID
(Rock)

6.

sion   sion   sion
   sion
   sion   sion
sion   sion   sion
   sion   sion

7.

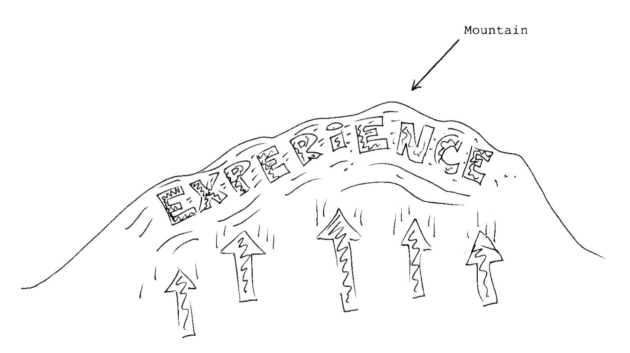

Earth Science 183

THE EARTH'S FORCES (continued)

3–2 Can You Solve the Problem?

Have students attempt to solve these eight brain-straining, mind-draining problems. Actually, few students will suffer any permanent damage since the activities are relatively painless.

1. There are three types of forces applied to solids within the Earth. Use the three groups of rhyming words as clues to these forces. Write the name of each force in the space provided.

Rhyming Words	*Force*
mention, pension, attention	_____
depression, regression, session	_____
steering, rearing, hearing	_____

2. The box contains clues to the answer to the question: What causes rocks deep within the earth to melt? Combine three different letter combinations to find the answer.

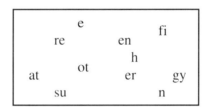

3. The stick figures represent the direction or application of force. Write the name of the force—tension, compression, or shearing—being illustrated in each sketch.

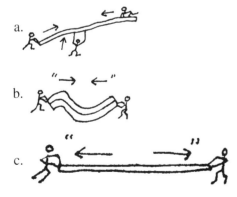

4. Various forces can keep the Earth's surface in a restless state. List at least one event due to force next to the letters that spell FORCE. The name of the event must begin with the letters that spell FORCE. The first one is done for you.

THE EARTH'S FORCES (continued)

F - folding, faulting
O
R
C
E

5. What are trapped in the box?

```
┌─────────────────────┐
│       Various       │
│  ces           ces  │
│  ces           ces  │
└─────────────────────┘
```

6. There are seven pairs of letters scattered inside the circle. If you put the paired letters together correctly, you will uncover two terms (beginning with the letter *f*) that describe the result of the Earth's internal forces on rocks. *Hint:* One term has six letters; the other term has eight letters.

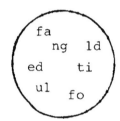

7. Use the two terms, stress and Earth, to show *how* the Earth is under stress.
8. This mystery force has two names. The first name has ten letters. A closer look reveals two *c*'s and two *"on's."* The second name has two *r*'s. Two vowels and three consonants remain. What is the name of the mystery force?

13–3 What's in a Name?

More names. More words. And more terms. Students need to reach deep into the think tank for clever answers to most of these items.

1. Make up rhyming names associated with the human body to match the listed terms related to the Earth's forces or the results of those forces. The first one is done for you.

    Example: heat - feet

    Strain   Twist
    Metamorphic   Tear
    Compression

THE EARTH'S FORCES (continued)

2. Diastrophism refers to any process responsible for deforming the Earth's crust and producing mountains and continents. Use the letters that spell DIASTROPHISM to identify *two* sections or structures of the human body. You may use a letter *only once*.

3. Three of the forces responsible for mountain building are pressure, heat, and tension. Use three letters from *pressure*, two from *heat*, and one from *tension* to spell the last name of a make-believe character. This character's first name, Les, rhymes with his last name. His strength produces deformation. Who is he?

4. Unscramble the letters in the four names to reveal the fault-block mountains located in Wyoming (two words) and in California (two words). These mountains are the result of a slipping/pushing force.

   DNGRA   NTTSOE
   RSRAEI   ASNDEVA

5. The arrows show the direction of force against the Earth's crust. Name the features that may occur from the direction of force.

   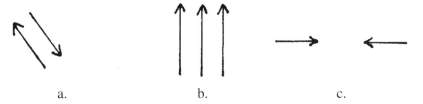

   a.   b.   c.

6. Another one of the Earth's forces is responsible for breaking down rocks and other materials by wind, water, and ice. What is the name of this force? To find out, use every letter in wind, water, and ice except one *i*, one *w*, the *c*, and the *d*. Then add an *h* and *g*.

THE EARTH'S FORCES (continued)

13–4 Create-A-Comment

Allow students to examine each drawing and write a comment or statement in the balloon above the figures in the sketch. Students should enjoy themselves by creating puns or gags related to the section theme.

1.

2.

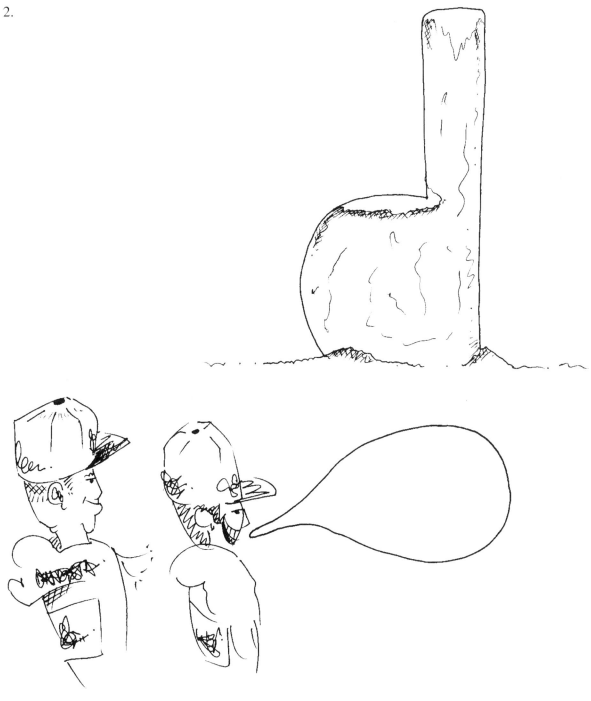

THE EARTH'S FORCES (continued)

13–5 Riddle Bits

Yes, more riddles for students to moan over. However, they do get excited when they discover a clever answer.

1. Where do you think these two items belong?
   a. What does the Earth's crust undergo when pressure builds?
   b. What force is synonymous with strain?
2. Weathering caused several rocks to roll down both sides of a hill. Somebody painted a different number on each rock. One day a person picked up rock numbers 1, 6, 8, and 9 and took them home. How many are left?

3. Janice and Nancy were carrying a tray of dishes. Suddenly the tray tipped over and the dishes crashed to the floor. Who is to blame? Use the sketch as a clue to the answer.

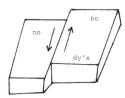

4. How can you show that pressure exerts an odor when it begins to disappear?
5. What do bath towels and the Appalachian Mountains have in common?
6. If sketch A represents uplifting, what does sketch B represent?

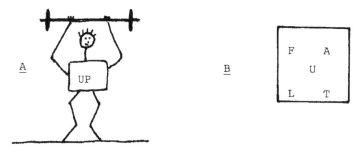

7. What do pinball machines and rock layers have in common?

Earth Science

THE EARTH'S FORCES (continued)

13–6 What Goes Where?

Let students accept the challenge of finding the right combination of symbols, words, letters, and numbers. If everything fits into place, they'll identify the correct terms. Remind them to unscramble the letters, symbols, and so on, and line everything up accordingly.

1. not out, but __ __ + g + not young + 6th letter.
2. "g" in the middle of two "moms."
3. old (minus d) + 10 + m.
4. [cloud with rain] + [golf tee "fore" (minus ee)] + 19th letter.
5. [castle] ← minus the 2 s's + tin (minus the *t*) + 7th letter.
6. 3 s's + [tree] + take away an *e*.
7. in + [corn] (3 letters) + 7th letter + sh.
8. symbol for silica + 10 + opposite of off.
9. f + symbol for lithium + opposite of down + t.
10. cl + [eye] + ine + [ant]

13–7 Creative Potpourri

This is a strange item section. Students who enjoy "playing" with ideas and approaching problems in a "tongue-in-cheek" manner will be right at home with these activities.

1. What does the following statement describe?

    "Colliding continental crusts cause crumpling."

2. Make up silly sentences using the listed words as key parts of each sentence.

    Example: Mary said, "This *stress* is one of my favorites."

    | tension | fault |
    | strain | shearing |
    | heat | thrust |

3. Four terms related to force fit the puzzle. What are they? Your only clues are the nine vowels in the puzzle.

THE EARTH'S FORCES (continued)

	e		i	o
		e	a	
		a		
			a	i
		i		

4. Complete each sentence or paragraph by finding the missing word. The missing word, of course, is related to the Earth's forces.

   a. If you can't stand the ___ ___ ___ ___ (four letters), then get out of the kitchen.

   b. Joyce's father put a lot of ___ ___ ___ ___ ___ ___ ___ ___ (eight letters) on her to earn top grades in school.

   c. Ron spent nearly all morning cleaning his room. His mother jokingly said, "I hope you didn't ___ ___ ___ ___ ___ ___ (six letters) yourself."

   d. Mrs. Marques returned home early from shopping. Her daughter, Louise, and four neighborhood children were watching television and being rowdy. You could feel the ___ ___ ___ ___ ___ ___ ___ (seven letters) in the air.

5. Fill in the blanks with the missing letters. Then answer the question.

   What do s__ie__t__s__s think might cause the m__v__me__t of lith__sph__r__c pl__t__s?

   *Answer:*_____

6. The listed words rhyme with parts of the earth affected by forces. Write the "Earth Words" in the spaces provided. The first one is done for you.

crate	_plate_	must	_____
motions	_____	shivers	_____
toil	_____	thrills	_____
fountains	_____	nation	_____
peaches	_____		

7. The box contains scrambled parts of five terms related to the Earth's forces. Unscramble the letters, put them together, and write the completed terms in the spaces below the box.

ta	noi	pu
moc	tl	ts
uaf	ar	ni
tfil	sserp	eh

   _____   _____
   _____   _____
   _____

Section 14

WEATHER AND CLIMATE

14-1 What Does the Sketch Represent?

Trying to decide what a sketch represents requires a student to be a flexible thinker, to employ an energetic imagination, and to possess the ability to accept a bizarre answer or two.

Ask students to study the sketch. Then have them write what they think the sketch represents on the line below the sketch.

1.

_____

2.

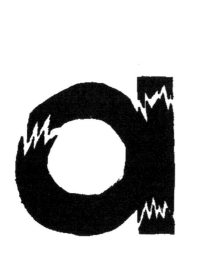 

3.

CYCLONES

4.

5.

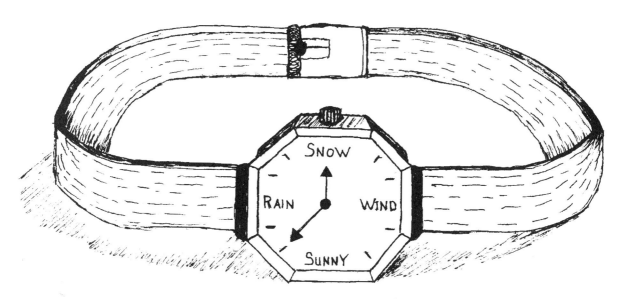

6.

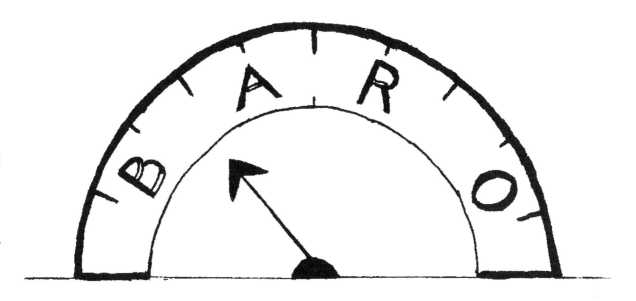

WEATHER AND CLIMATE (continued)

14–2 Can You Solve the Problem?

Problems, problems, and more problems. What a way to start a class period! Happily, students at all academic levels find them challenging and fun. Here are some wonderful stimulators to get you started:

1. Unscramble and combine the chemical symbol of each element to form a word that relates in some way to *weather*. The first one is done for you.

    Example: nitrogen, sulfur, and uranium—n, s, u—*sun*

    a. iridium and argon (remove an "r")

    b. nitrogen, iodine, and radium

    c. oxygen, tin, and tungsten

    d. sulfur and galium

2. Precipitation is water that falls to Earth's surface from the atmosphere. The scattered letters in the box make up the names of four different types of precipitation. Use each letter once to write the names in the spaces below the box.

    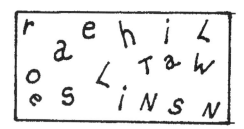

    a. _____
    b. _____
    c. _____
    d. _____

3. Use the letters in the word PRECIPITATION to write the names of 12 items that might be found *in* or *on* a kitchen refrigerator. You may use a letter more than once.

4. Prove or disprove this statement:

    There is a way to get *moisture* out of *the atmosphere*.

5. Climate is long-term weather; that is, the average of all weather conditions of an area over a period of years. List seven objects found around your school that are affected by climate. Each object must begin with the letter designated to the left of the space.

Earth Science

WEATHER AND CLIMATE (continued)

c _____
l _____
i _____
m _____
a _____
t _____
e _____

6. You are watching the evening news. At the end of the program, Betty Breeze, local meteorologist, flashes several lines of information on the screen. What is she showing you?

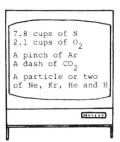

14–3 What's in a Name?

Name droppers will love this activity. These starters take several minutes to complete and give creative neurons room to expand. In some cases, students working in pairs experience a higher success rate than students working alone.

1. Make up rhyming first names of boys and/or girls from the weather terms. You may abbreviate names; however, do not use nicknames. The first one is done for you.

    Example: storm—Norm Storm

    snow   high
    rain   low
    hail   sleet
    dew   ice

2. Let your imagination run wild. See how many first and last names you can create out of the weather and climate terms. The first one is done for you.

WEATHER AND CLIMATE (continued)

Example: anemometer—Anna Mometer

millibar	radiosonde
aerosals	maritime
radiation	cyclone
aneroid	hurricane
ionosphere	altostratus
Fahrenheit	typhoon

3. List a hobby or occupation that matches the first and last names you created in Exercise 2. For example, an anemometer measures wind velocity. Therefore, anemometer or *Anna Mometer* might be the name of the person who records wind velocity for a local weather station.

4. Some weather and climate terms have the same letter repeated two or three times. See how many of the "repeaters" you can find.
   a. *Four* terms with *two* i's.
   b. *One* term with *three* o's.
   c. *Four* terms with *two* a's.
   d. *Five* terms with *two* o's.
   e. *Three* terms with *two* r's.

5. See if you can list ten different nouns from the letters in the term METEOROLOGY. You may use a letter more than once.

WEATHER AND CLIMATE (continued)

14–4 Create-A-Comment

Make way for some cartoon fun. In this activity students develop their own thoughts on how an illustration might lend humor to a situation.

Provide students with a copy of a drawing. Allow students to think about the drawing for a minute or two. Then ask them to write a comment or statement in the balloon above the two figures in the illustration. The statement, of course, should relate to the theme.

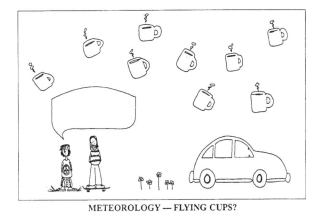

METEOROLOGY — FLYING CUPS?

METEOROLOGY — IT MAKES CENTS.

WEATHER AND CLIMATE (continued)

14–5 Riddle Bits

These items do more than test a student's ability to think creatively; they pry loose a groan or two. Regardless, they set the cerebral wheels spinning. So let's give it a whirl.

1. What kind of bus do you usually see during a rainstorm?
2. Why is a rabbit a good indicator of humidity?
3. What weather term is made up of two boy names? *Hint:* Wind is moving air.
4. What atmospheric gas is this?

5. This is what you run into as you climb in altitude. What is it?

    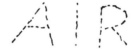

6. What pressure belt shows the greatest percussion effect?
7. If this symbol (⊔────) shows wind speed and direction on a weather map, then what does this symbol indicate? (w e a t↙h e r)
8. Melissa said to Francine, "I'll give you my wind for your wind." Francine replied, "Okay, it's a deal!" What atmospheric event were these girls demonstrating?
9. What well-known pressure belt carries an animal's name?
10. Which of these weather terms suggest something very noisy—thunder, lightning, or a cloud?
11. What does the underlined word indicate?

    latitudes   latitudes   <u>latitudes</u>   latitudes   latitudes

12. If you could tell the occupations of people by what they do during their leisure time, then what do you think a person who yells "Fore" before tossing a fishing lure into a stream does for a living?
13. What do these names indicate?

    Aunt Joan Humidity
    Uncle Bill Humidity
    Cousin Mary Humidity

Earth Science

WEATHER AND CLIMATE (continued)

14. Typical weather map symbols are as follows: (R) rain (S) Snow (F) Fog. What might these symbols on a weather map indicate?

15. What kind of weather is happening in Australia?

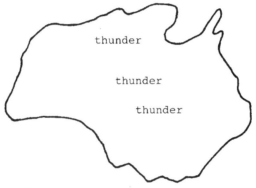

14–6 What Goes Where?

This bagful of symbols, words, and numbers is guaranteed to push a few creative buttons. Simply ask students to identify the symbols, decipher the words and phrases, and organize everything to reveal the correct term.

1. **1** + Wizard of _ _.

2. Class of birds + w.

3. g + 💡 + opposite of *out* + n.

4. Two "on's" + o + *m*aritime.

5. detar (backward) + 🪐 − rn.

6. 👁 + sicklike + 🏢 vertical steel rod or _ _ _ _ + m

7. ☀ + 🚢 (Not ocean, but _ _ _)

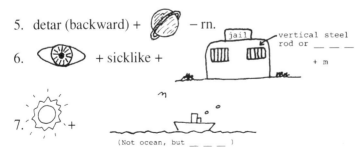

201

WEATHER AND CLIMATE (continued)

14–7 Creative Potpourri

These activities—a mixture of puns, crazy definitions, weird sentences, and miscellaneous "fun stuff"—allow students to go on a creative rampage. And perhaps best of all, they act as motivators for students to develop their own literary fun.

1. Test your creative patience with a pun or two. A pun plays upon words similar in sound, but having a different meaning.

Read the following pun statements and record what you think they mean.

   A. Did you hear about the meteorologist who was fired for not getting cirrus?

   B. Torricelli, the Italian scientist, had little trouble working under pressure.

2. Weather proverbs may be based on careful observations or superstition.

Read the two "proverbs." Then record if you think they are based on fact or fiction. Use any evidence you wish to support your answers.

   A. When the cirrus starts to form
   Prepare to face a freezing storm.

   B. When the wind blows from the west,
   The fish will bite their very best.

3. Use the weather and climate terms to write weird sentences. For example, the word *anemometer* might look like this:

   Anemometer teacher at Open House last night.
   (Anna's mom met her teacher at Open House last night.)

Here's another example:

   "Steve, you can barometer stick from Joan."
   ("Steve, you can borrow a meter stick from Joan.")

moisture	polar
snow	doldrums
thermal	maritime
wind	hurricane
solar	pressure

Earth Science

WEATHER AND CLIMATE (continued)

4. Make up silly, "scientific" definitions for the weather and climate terms. The first two are done for you.

   Atmosphere—A phobic condition brought on by the prefix "atmos."
   Dew point—The time to begin a task or project.

doldrums	overcast
cyclone	humid
gases	molecules

5. What do you think the weather terms listed in group A have in common with the words listed in group B? Record your answers in brief sentences.

Group A	*Group B*
a. mass	a. mother
b. condensation	b. lion
c. freeze	c. sneeze
d. current	d. apartment
e. isotherm	e. smother

6. Make up two or three personalized car license plates relating to weather or climate. Here are two examples:

   ⬚ 4 CSTR ⬚   ⬚ DR THNDR ⬚

Section 15

ASTRONOMY

15–1 What Does the Sketch Represent?

One way to kick start the imagination is to look at an illustration and decide how it fits a theme or subject. Have your students activate their neurons with the following artistic stimuli. After they study each sketch, ask them to write what they think the sketch represents on the line below the sketch.

1.

2.

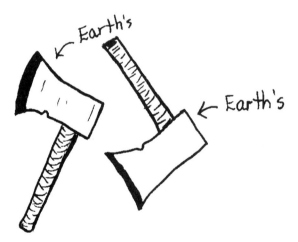

3.

4.

5.

6.

7.

Earth Science

ASTRONOMY (continued)

15–2 Can You Solve the Problem?

Miniproblems with an astronomy theme help anxious students stay busy. Keep students mentally active by providing them with any of the seven items. Exercises 6 and 7 require students to think in "riddle terms."

1. Find the names of *four* planets in the word PLANETARIUMS. You may use a letter more than once.

2. Some space objects revolve around the sun. The letters that identify *four* such objects are scattered around the sketch. Unscramble the letters and discover what they are.

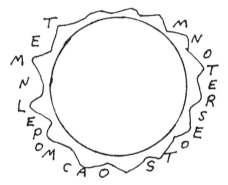

3. As you know, the sun has a circular shape like the letter O. List *three* terms related to the sun that have at least *two* O's in each term.

4. The sun has been known to produce what is being shown by the scattered letters. Unscramble the letters, put them together, and you'll know the answer.

5. Pluto takes about 250 Earth-years to circle the sun once. How many earth-days would this be?

ASTRONOMY (continued)

6. You're reading a story about eclipses. You run across this word: ECLIP. How do you think the word relates to the theme of the story?

7. The east wall of Professor Luney's room looks like this:

ISOLARI	ISOLAR I	ISOLAR I	ISOLAR I	ISOLAR I
ISOLARI	ISOLAR I	ISOLAR I	ISOLAR I	ISOLAR I
ISOLARI	ISOLAR I	ISOLAR I	ISOLAR I	ISOLAR I
ISOLAR I	ISOLAR I	ISOLAR I	ISOLAR I	ISOLAR I

Professor Luney, of course, teaches astronomy at the local university. What kind of covering decorates the wall?

15–3 What's in a Name?

The five miniactivities provide students with an opportunity to identify names from astronomy terms by moving letters around or filling empty spaces.

1. Find the missing letters in each term. *Hint:* The missing letters help complete the first name of a girl.

   A _ _ U L A R   S _ _ A R
   S _ _ L L _ R   P _ _ N _ T
   C _ R O N _   R A _ A R

2. Find terms that rhyme with the first names of girls and boys. Write them next to the girl or boy name.

   Example: Janet Planet

   Sally _____   Janice _____
   Mona _____   Mace _____
   Lars _____   Dave _____
   Michael _____   Brian _____
   Nora _____   June _____

3. Match the space terms with rhyming words that identify parts of the human body.

   Example: solar molar

   sky _____   flare _____
   space _____   eclipse _____
   apogee _____   rocket _____
   heat _____   gravity _____

Earth Science

ASTRONOMY (continued)

4. Match the animal names with rhyming astronomy terms. Write them next to each name or names.

   Example: antelope telescope

   fly   _____   dalmation   _____
   loon   _____   zebra   _____
   mite   _____   polar bear _____ (two words)
   gar   _____   musk Ox   _____ (one word)

5. Add a letter to the beginning of the word or prefix to change its meaning. Write the new term in the space provided. Use the clues given in parentheses. *Rule:* The changed version must relate to science or health.

   astro _____ (stomach)   ursa _____ (joints)
   light _____ (disease)   air   _____ (body cover)

ASTRONOMY (continued)

15–4 Create-A-Comment

Give students copies of the two sketches. Have them write what they think would be appropriate comments or statements in the balloons above the figures in the illustrations.

Ask students to keep it fun by creating humor tailored to the section theme.

1.

2.

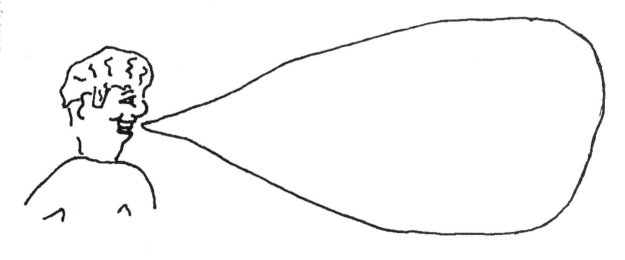

ASTRONOMY (continued)

15–5 Riddle Bits

The following 14 items will test the patience and flexible thinking ability of each student. This is an excellent section for students who like riddles and don't mind "far-out" humor.

1. What make of television set shows the highest number of space programs?
2. What do astronomers on horseback use to see at night?
3. What planet has an "eye" but cannot see?
4. Why are there no flies on Venus?
5. What name of a space object sounds like the noise a cow makes?
6. What planet gives the best indication of air pressure?
7. What planet is known for the music it makes?
8. What planet is known as the "go left, go right" planet?
9. What giant star in the constellation Scorpius has an insect as part of its name?
10. What is in the middle of the largest planet?
11. What would you be if a total eclipse occurred and you were trapped in the middle of a penumbra?
12. What planet wears a hat?
13. What are these seven groups of make-believe constellations?

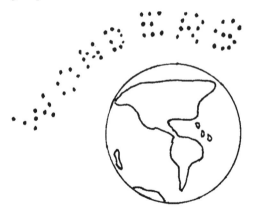

14. Why is Saturn known as the "bathtub planet"?

15–6 What Goes Where?

Students can sharpen their creative thinking skills by unscrambling the mixture of symbols, words, and numbers to reveal ten mystery terms.

Earth Science   213

ASTRONOMY (continued)

Example: E T + [rock] = E T + ROCK or ROCKET.

1. E + [lips] + E L.
2. [eye] + Roses are red, violets are blue + · · · + U N.
3. [mouse] + E R + C.
4. ½ + R E.
5. S + [arrow] + A.
6. A R + E S + [ant].
7. [bed] + O + A L.
8. *Hint:* A star (two words).

   Not pa, but _ _ + P + [beans] + I (first word).

   R I + [1¢ coin] + symbol for gold (second word).

9. Veterans Administration (abbr.) + opposite of yes.
10. T + [rabbit] + H.

15–7 Creative Potpourri

Allow students to tackle a pun or two or any other item designed to push their creative buttons.

1. How about a trip to Punsville? A pun plays upon words similar in sound, but having different meaning.

Read the following pun statements and record what you think they mean.

   a. The sun with its great outbursts gained prominence with early astronomers.
   b. Jupiter is the most jovial planet in the solar system.
   c. A fantastic book has just hit the market entitled *Reflections from a Newtonian Telescope*.

2. List *eight* phenomena related to the sun. Begin each item with the letter *s*.

   Short
   Regular
   Long

ASTRONOMY (continued)

3. List *eight* names of automobiles that have names common to astronomy.
    Example: Vega

4. Briefly describe how you think these statements relate to astronomy.

    "Donna's ideas are cosmic."

    "George's chances to date Lora are light-years away."

    "Melanie has stars in her eyes."

    "Many wealthy people pay astronomical prices for their homes."

5. Combine two rhyming words that relate to astronomy. Try to create at least four of them.
    Example: Mars/stars

Section 16

OCEANOGRAPHY

16–1 What Does the Sketch Represent?

This section will provide eight sketches for the imaginative learner. Ask students to study the sketch. Then have them write what they think the drawing represents on the line below the sketch.

1.

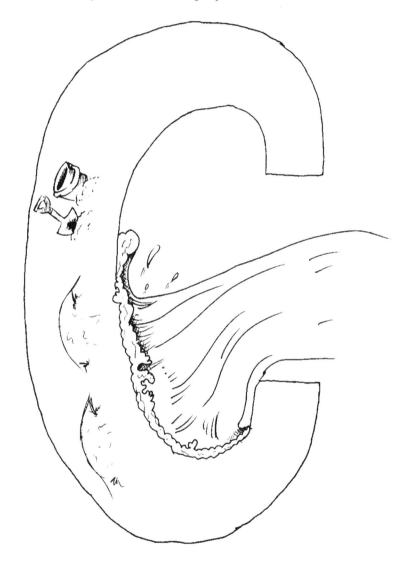

215

2.

3.

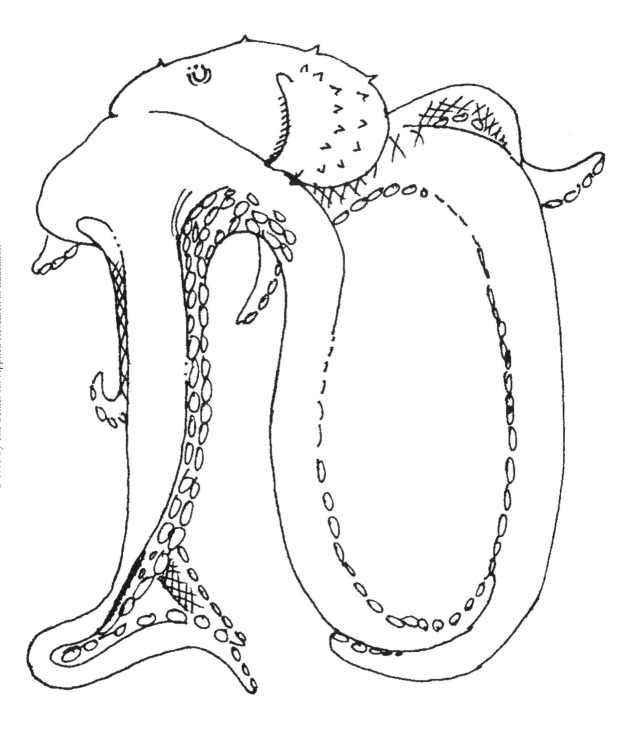

4.

5.

Hint: Two male fish from the same family

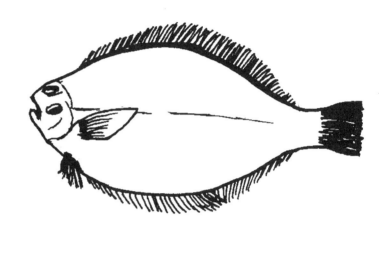

6.

7.

OCEANOGRAPHY (continued)

16–2 Can You Solve the Problem?

These miniproblems are excellent motivators that can be used at any time—as homework assignments, at the end of the period, as part of labs, and even in the middle of the period.

1. You can use the letters *only once* to spell the names of five common tide pool organisms.

2. How many *tuna fish* did the shark eat? Use the letter clues to help you find the answer.

3. The Caribbean Sea lies north of South America. If the sea in Caribbean disappeared, what would be left?

4. Show, by connecting letters with lines, how to make the word COD at least eight times by using three C's, four O's, and two D's.

         C   O   D
         O   C   O
         D   O   C

5. A story is told of a boy whose first name was *Neal*. He had an interesting last name. In fact, his last name was the same as the name of a benthos organism. What was it? *Hint:* His first *and* last names come from the letters that spell *barnacle*.

6. Ocean sediments are formed by materials that settle on the ocean floor. They may come from

Earth Science 223

OCEANOGRAPHY (continued)

river deposits, remains of living organisms, and materials from volcanic eruptions. List eight different sediments. Use the letters that spell SEDIMENT as the first letter in each item. The first one is done for you.

S - sand, salt, shale, and slate

E -

D -

I -

M -

E -

N -

T -

S -

7. *What* are these lumps found on the ocean floor and *what* are their mineral content. *Hint:* Unscramble the letters.

Fe	Fe		Ni		Ni		Mn			Mn	Fe	Ni
Fe		Fe	Ni		Ni		Mn			Mn		
Fe		Fe	Ni		Ni		Mn			Mn	Fe	
Fe		Fe	Ni		Ni		Mn			Ni		
Fe	Fe		Ni	Ni	Ni	Ni	Mn	Mn	Mn	Fe	Ni	Mn

Fe	Fe	Fe	Fe	Ni	Fe	Mn	Mn		Ni		
Fe			Fe	Fe			Fe	Ni	Fe		
Fe			Fe	Mn	Ni	Fe	Mn		Fe	Fe	
Fe			Fe			Mn	Ni		Ni	Mn	
Fe	Fe	Fe	Fe	Mn	Ni	Fe	Fe		Ni		

8. Connect the two groups of letters with a line to form words that identify each sketch. Write the completed word in the empty space. Now connect the completed word and sketch with a line.

		Completed Word
yot	ben	_____
nod	ton	_____
thos	gu	_____
nek	ules	_____

OCEANOGRAPHY (continued)

9. Find *four* nekton organisms hidden in the letters:

   dindolphincocsquidacrabphywhalestrturtleodoyster

10. Find *four* benthos organisms hidden in the letters:

    tospongelcodtzcoralapanemonerusharkmaoystersnjellyfish

11. Find *two* plankton organisms hidden in the letters:

    diatdinoflagellatesinradiolariansfishphyseahorse

12. Darken the spaces in the box where letters can be found that spell the answer to the question. A letter may appear several times.
    a. What large body of salt water covers more than three-fifths of the Earth?
    b. What sea animal is revealed in the puzzle?

    | b | d | f | u | i | m | r | s | p | i | m | k | g |
    | g | i | n | a | c | p | u | h | n | c | a | t | i |
    | f | e | r | m | l | c | k | e | s | r | l | a | p |
    | o | w | h | f | r | g | o | u | i | w | r | d | o |

13. See if you can find *five* words that identify sea organisms or relate in some way to oceanography. However, the words *must* contain the same double vowel (ii, oo, ee, aa, and so forth).

    Example: *oo*ze

14. See if you can find *five* words that identify sea organisms or relate in some way to oceanography. However, the words *must* contain the same double consonant (tt, ff, gg, cc, etc.).

    Example: he*rr*ing

16–3 What's in a Name?

This section includes six miniactivities that will test the creative spirit of students.

1. Read the nonsensical paragraph. Now rewrite the paragraph and replace the five underlined words with sea organisms whose names rhyme with each of the underlined words. Reread the paragraph. What a difference!

   Scales and *plarks* are well-known ocean critters. Smaller creatures such as *sweils* and *snerch* swim in the same environment. And, of course, Mr. *Slab* scurries about the sandy bottom.

Earth Science

OCEANOGRAPHY (continued)

2. Use the following words in "silly sentences." The first one is done for you.

   Example: sonar - Greg knew that *sonar* or later he'd have to take the test.

atoll	buoy	nodule
saline	tuna	sand
basin	island	beach

3. Match the words with items commonly found around the kitchen or garage. Be sure the word/item combination rhymes.

   Examples: coast - toast, coast - roast, coast - host

shore	whale
trench	beach
sea	jetty
reef	clam
fish	boat

4. One day a strange person was heard babbling to himself as he walked along the beach. He muttered about having supernatural powers and the ability to do great things. He was covered with a slimy mud. What would be a fitting name for this individual? (Think about ocean sediments.)

5. See how many *verbs* you can make out of the letters in the word OCEANOGRAPHY. You may use a letter no more than two times.

6. See how many girl's names you can make out of the letters in OCEAN. You may use a letter more than once.

225

OCEANOGRAPHY (continued)

16–4 Create-A-Comment

Hand out copies of the following drawings. Ask students to think about each drawing and then write a comment or statement in the balloon above the figures in the illustration.

Encourage students to write funny comments or statements. Give them freedom to create puns or gags related to the section theme.

1.

OCEANOGRAPHY (continued)

16–5 Riddle Bits

Turn your students loose and let them wade through a few riddles. The key to solving these riddles lies in flexible thinking and the patience to handle a dab of "punomasia."

1. There are ctenoid, ganoid, and placoid fish scales. What kind of scales appear on this fish?

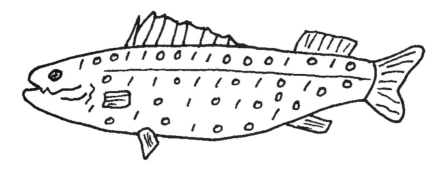

2. What would you get if you crossed a . . .
   starfish with a sea urchin?
   whale with a sole?
   shark with a clam?
   perch with a smelt?
   whale with a scallop?
   squid with a shark?
   whelk with an eel?
   flounder with a hake?
3. What vegetable can be found in a tropical sea?
4. Why did the commercial fisherman constantly say "Huh?"
5. What is the most verbal animal in the sea?
6. Why did the guy bury a dollar bill in the beach sand and return several hours later to dig it up?
7. What two organisms could work together as "Carpenters of the Sea?"
8. What seagoing vessel carries the name of the upper portion of the thigh?
9. What part of a sea wave is the most peaceful?
10. What part of a sea wave is most violent?

OCEANOGRAPHY (continued)

16–6 What Goes Where?

Give students the following set of symbols, words, and numbers to unscramble and reveal the mystery terms.

1. [die showing 6] (minus *e*) + [atom symbol]

2. [head] + 2,000 pounds.

3. Short for Lester + du + not yes, but ___ ___.

4. [fish] + 2 + <u>nature</u>.

5. C + C + C + C + C + C + C.

6. "C" "C" "C" "C".

7. *An added challenge for your students:* How many symbol/letter combinations can you create from these terms: breaker, porpoise, shoreline, and electric eel.

    Example: shoreline - shhh + straight _____ + e + or.

16–7 Creative Potpourri

The open-ended activities in this section provide a humorous challenge for students who enjoy literary freedom. There are no right or wrong answers, only responses that students create to match each item.

1. List *five* things related to oceanography that begin with the letter *w*.

2. *Coral Graffiti*

   If you were a sea creature and liked to write graffiti, what clever things might you produce? In the open areas on the coral reef, write two or three graffiti messages about fellow sea creatures.

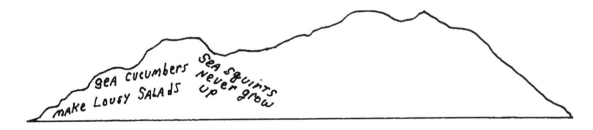

OCEANOGRAPHY (continued)

3. List *six* canned products from the sea carried by supermarkets on their shelves.
4. Creatures of the sea find their way into expressions or statements people make. For example, you might hear somebody refer to another person as an "old crab."

Make up three or four expressions or statements using various sea critters as the main character.

5. The "reef creatures" is a "gang" made up of sea organisms. There are eight "gang" members—squid, crab, clam, fish, eel, shark, shrimp, and worm. Create a fun name for each organism. As an example, Sid the Squid.
6. Write the titles of three or four Hollywood movies about scary sea creatures or monsters. Identify the main character in the movie.

    Example: Jaws - main character: great white shark.

7. Create your own titles for three or four make-believe movies that use sea creatures or monsters as main characters.

LIFE SCIENCE

- ▼ SIMPLE LIFE
- ▼ PLANT LIFE
- ▼ ANIMALS WITHOUT BACKBONES
- ▼ ANIMALS WITH BACKBONES
- ▼ REPRODUCTION AND DEVELOPMENT
- ▼ GENETICS AND CHANGE
- ▼ HUMAN BIOLOGY
- ▼ HEALTH AND ENVIRONMENT

Section 17

SIMPLE LIFE

17–1 What Does the Sketch Represent?

A picture may be worth a thousand words but finding a matching description for a sketch often stretches the brain cells to their limit.

The following eight activities offer a terrific way to begin a unit. Let students select a sketch or two and write what they think the sketch represents on the line below each drawing.

1.

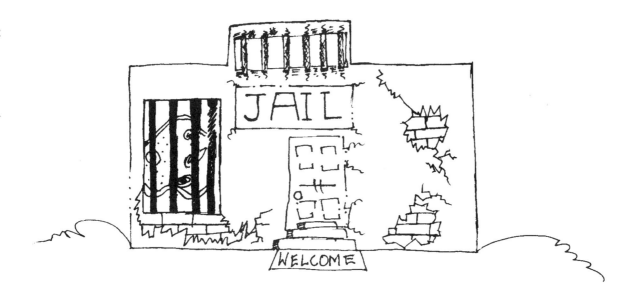

2.

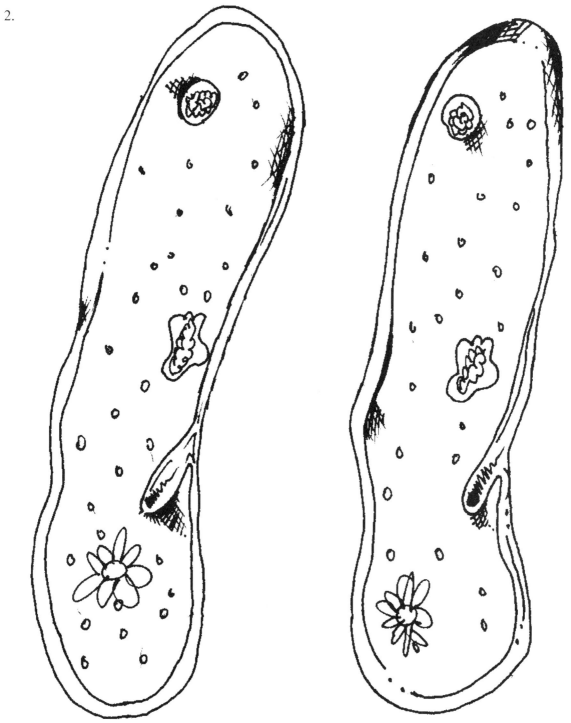

3.

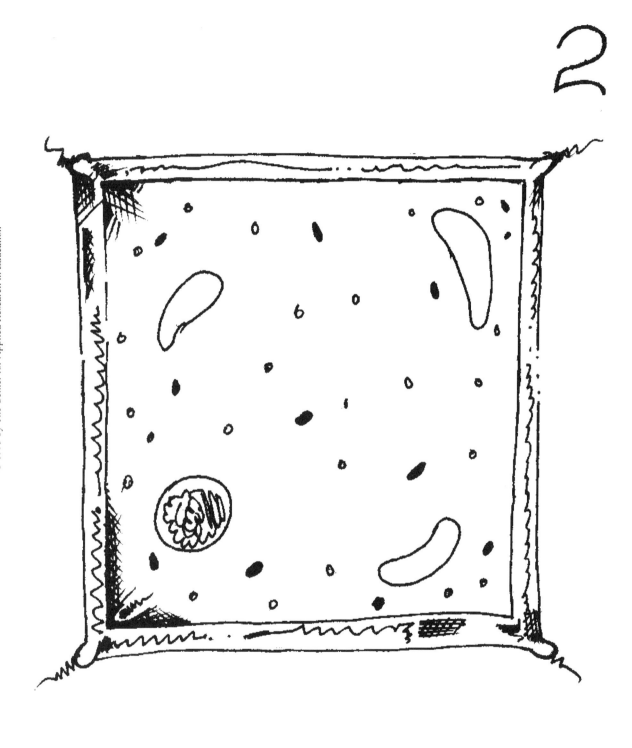

4.

5.

6.

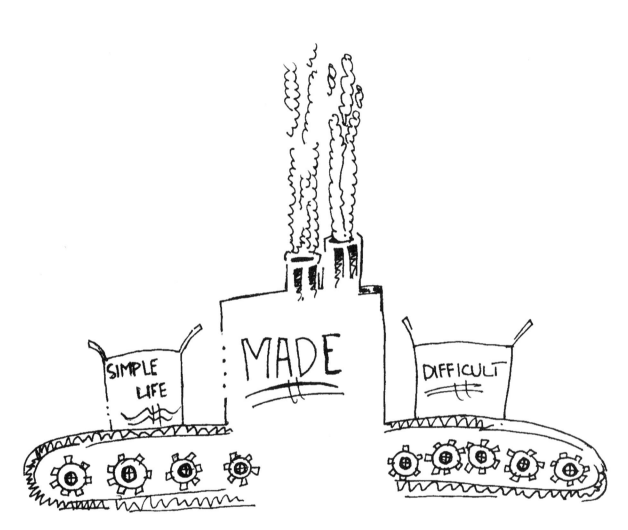

7.

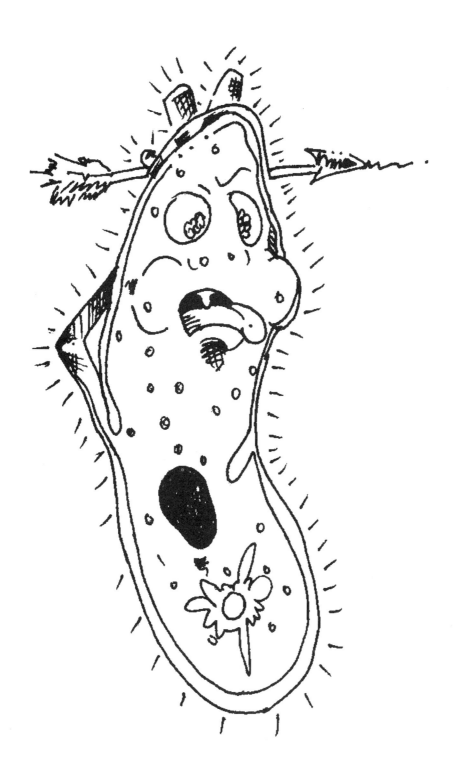

8.

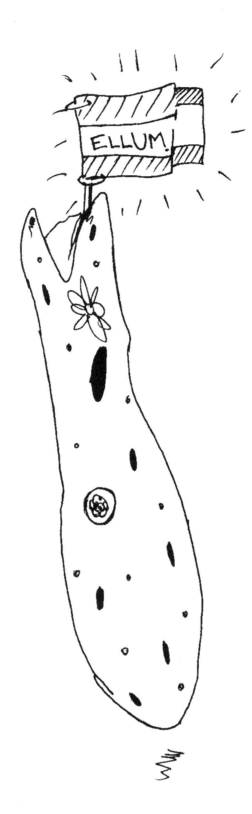

Life Science

SIMPLE LIFE (continued)

17–2 Can You Solve the Problem?

Simple life sometimes produces difficult problems. However, this section won't bruise the brain too much, only titillate a creative neuron into service.

1. Letters that make up two simple plant names are scattered in the sketch. Four letters, two for each plant, are missing. Find the missing letters, put them with the letters in the sketch, and complete the plant names.

2. People keep calling Mrs. X by the wrong last name. They call her Mrs. Lenguae or Lugeena or Nalguee or some other weird name.

Mrs. X has three children, a daughter and two sons. She named them from the letters that spell her *correct* last name. One boy is called Len. What are the names of the remaining children? *Hint:* Her last name is the same as a one-celled, microscopic animal.

3. Bacteria are simple, single-celled organisms belonging to the Moneran Kingdom. The names for the three shapes of bacteria appear in the sketches. Unfortunately, the letters are scrambled and may or may not match the correct sketch. Unscramble the letters and write the name of the bacteria's shape in the blank sketch that matches the description.

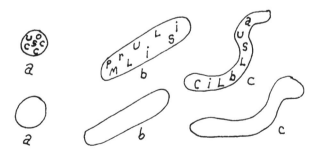

SIMPLE LIFE (continued)

4. Phylum: Protozoa. Class: Mastigophora. Order: Phytomonadina. Genus: Chlamydomoas. What a bunch of tough names for a single-celled, microscopic organism!

Would a simple description give a clear picture of what a *Chlamydomas* looks like? Let's find out. Read the description carefully. Then sketch how you think the organism appears.

Description

- One end of organism four times wider than opposite end
- Two flagella attached to narrow end
- Hairlike structures line inside wall of large end
- Two horizontal, dotlike structures located at narrow end

5. Show a way to reduce the amount of green, brown, or red algae by taking something away.

17–3 What's in a Name?

Names. Words. Terms. The following six miniactivities will provide students excellent practice in getting to know them.

1. Create first and last names out of the following terms related to simple life.

    Example: algae - Al Gee.

As in the example, you may want to change a letter or two.

diatoms	pseudopod
amoeba	rotifera
budding	decomposer
virus	rickettsia
Moneran	

2. Mr. Slime Mold claims you can establish his age and body shape by studying his name. Go ahead and see if he's telling the truth.
3. Can you uncover the name of the mystery organism? Maybe these two clues will help:
   - It's two different organisms in one.
   - A crusty, colored patch on rocks.
4. Vorticella is a name of a microscopic organism. *Show*, by using a line, that it is a one-celled ciliate.
5. A dinoflagellate is a one-celled protist. If it takes three of them stacked end to end to form the letter *d*, how many would it take to make *DINOFLAGELLATE*?
6. List *five* simple organisms whose names begin with the letter *r*.

Life Science   243

SIMPLE LIFE (continued)

17–4 Create-A-Comment

Students will enjoy grappling for a comment or statement to match the theme of the illustration: amoeba humor.

Provide students with a copy of the drawing. Have them study the illustration and write a comment or statement in the balloon above the figures in the drawing. The more creative students will do both sketches.

1.

2.

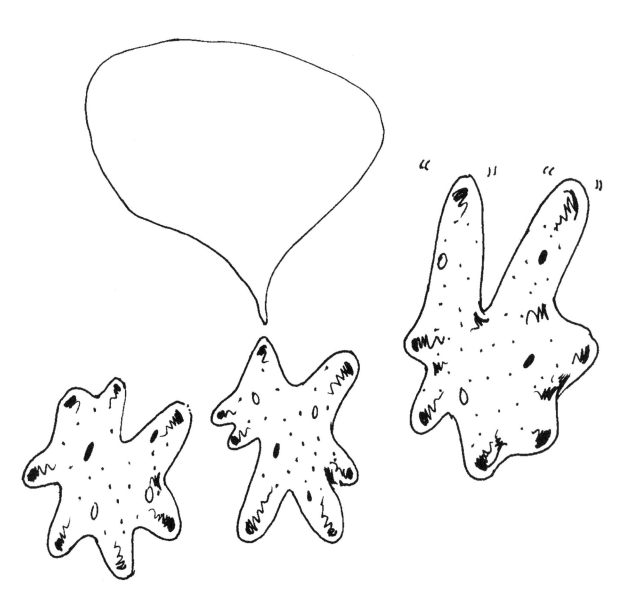

Life Science   245

SIMPLE LIFE (continued)

17–5 Riddle Bits

It's not easy trying to solve a riddle. Most riddles require the brain cells to remain open and receptive to imaginative stimuli. Here's an opportunity for students to go on a creative thinking spree.

1. What simple, flowerless organism is half plant and half female domestic bird?
2. How is a salmon like a mushroom?
3. What would happen if the sun became covered with bread dough?
4. What could we call our world if it housed only diatom organisms?
5. A scientist trapped 25 plankton in his net. He weighed them. How much did they weigh?
6. Mrs. Tista, an expert research scientist, spends most of her time studying microorganisms. She is highly respected by her colleagues. What do they consider her to be?
7. What do you call identical amoebas?
8. How do amoebas communicate?
9. What do you call a stupid fungus?
10. If a mold is mostly old (m*old*), what is a rotifer?
11. What is featured in the sketch?

euglena

12. Mr. Mushroom, Ms. Puffball, Mr. Smut, and Mrs. Rust are members of the same social organization. What is the name of the organization?

17–6 What Goes Where?

Have students rearrange the symbols, letters, and numbers to make sense out of the jumbled series. Then, if everything goes well, they'll identify the correct term.

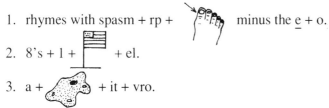

1. rhymes with spasm + rp + [foot image] minus the e + o.
2. 8's + 1 + [flag image] + el.
3. a + [amoeba image] + it + vro.

SIMPLE LIFE (continued)

4. air + [stick figure "to be or not to be..."] + ab + e.

5. le + opposite of old + 3rd letter + su.

6. [eye] + r + su + 22nd letter.

7. opposite of off + [fish] + "aye."

17-7 Creative Potpourri

This section provides a mixture of activities—identifying sketches, miniproblems, finding the correct terms, and locating the mystery words. These activities are designed for the flexible thinker.

1. Look at the sketch. Write down the first thought as to what the illustration means.

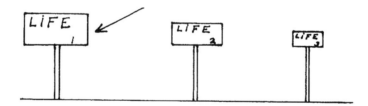

2. Sarcodines are organisms who use pseudopodia for locomotion. The amoeba is a common freshwater sarcodine. What saltwater fish "hides out" with sarcodines?

3. Look at the four columns of jumbled words, parts of words, exclamations, and so on. Two of them in each column fit together and sound like the name of a simple plant or animal. Write the name of the plant or animal in the space under the column.

A.	yah	B.	Tom	
	sill		die	
	ya		Adam	
	yachts		tum	
	silly		diet	
	_____		_____	
C.	ham	D.	Ken	
	if		skin	
	four		like	
	men		la	
	fura		light	
	_____		_____	

Life Science

SIMPLE LIFE (continued)

4. Use the letters in the word SPIROCHETE to form words that spell the answers to each listed question. You may use a letter more than once.

   A. Fungi reproduce when certain units develop into new organisms. What are these units called?
   B. What substance composes the cell walls of most fungi?
   C. Under what heading do scientists classify microscopic organisms that are neither plant nor animals?
   D. What is the name for spherical or egg-shaped bacteria?

5. Look at the sketch. Write down the first thought as to what the illustration means.

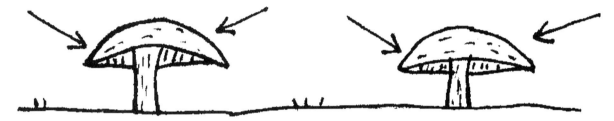

6. There are three mystery words hidden in the four rows of letters. Use the clues to help you locate the words. Circle the words.

   Clues

   A. I am trapped in an animal cell.
   B. I give viruses their shapes.
   C. I give the paramecium its slipperlike shape.

Note: Answers may be forward and backward.

```
p r i o n e l o i r t n e c r o t i f e r a
s u r i v c a p s i d s e u g l e n a a n d
e l c i l l e p o r g a n i s m s d l o m e
o r s a p r o p h y t e a n d i l l i c a b
```

7. It's pun time. A pun plays upon words similar in sound, but having a different meaning. Read the following pun statements and record what you think they mean.

   a. Did you hear about the mycologist who took a lichen to work?
   b. And then there was the unscrupulous mushroom collector who had no morels.

Section 18

PLANT LIFE

18-1 What Does the Sketch Represent?

These sketches allow students to think freely, play with words, and produce humorous responses. Have students examine a sketch and write what they think the illustration represents on the line below the drawing.

1.

2.

3.

4.

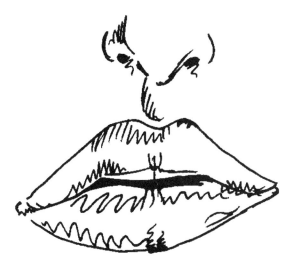 

5.

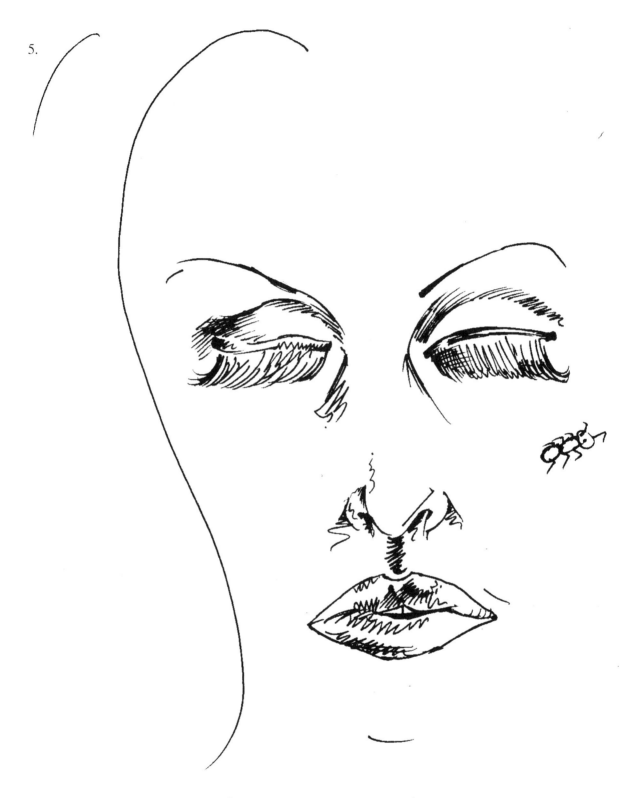

6.

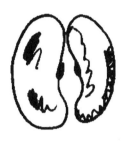

one cotyledon
two cotyledons

three cotyledons
four cotyledons

five cotyledons
six cotyledons

7.

Life Science   255

PLANT LIFE (continued)

18–2 Can You Solve the Problem?

This section should be titled "Brain Bruisers." The seven miniproblems require students to think rationally, yet remain open for unique responses, that is, slightly offbeat answers.

1. A fictitious Spanish scientist, Dr. Joseph Hidalgo, claims you can find gold (Au) in chlorophyll, the green pigment in plant cells needed for photosynthesis. How could this be true?

2. *Show* three divisions of bryophytes.

3. Horsetails are among the first tracheophytes (plants in which water and nutrients move through tubelike cells) to appear on earth. They look like this:

    Sketch a horsetail about 3 feet from the ground.

4. Tendrils are slender projections from the stem that are used for support by some plants. *Show* how tendrils might appear around the stem shown in the box.

5. A rhizome is an underground stem that can produce new plants. Swollen areas at the end of a rhizome are called tubers. A white potato, for example, is a tuber. How many "bers" can you count?

6. There are *four* plant tissue structures trapped in the leaf diagrams. As you can see, their names are separated into 12 letter groups. Join the letter groups to form each of the three plant tissues. Write their completed names in the spaces to the right of the diagrams.

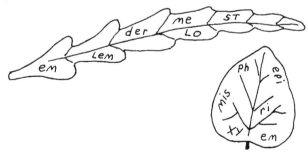

PLANT LIFE (continued)

7. Here are 12 scattered letters. If you put the correct 7 letters together, you'll have the answer to the following question: What are leaves collectively called?

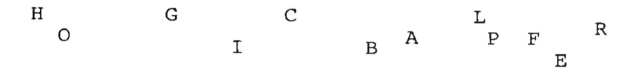

18–3 What's in a Name?

This section combines various words, terms, and names to test a student's ability to find a satisfactory answer. The six activities pose a modest challenge and keep students working at a steady pace.

1. Some plant life terms have the same letter repeated two or three times. Find examples for each of the listed letters.

   a. *Three* terms with *two* a's
   b. *Four* terms with *two* i's
   c. *Two* terms with *three* o's
   d. *Three* terms with *three* e's
   e. *Four* terms with *two* t's

2. Fill in the spaces with letters completing the terms related to *plant structures*.

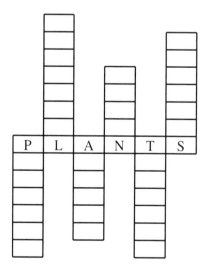

Life Science

PLANT LIFE (continued)

3. Epidermis is the outermost layer of cells of an organism. List ten three-lettered words from the letters in EPIDERMIS. You may use a letter more than once. However, do not list abbreviations, prefixes, names of people, or nicknames.

4. Make up girls' first and last names from the listed terms. The first one is done for you.

Term	Girls' First/Last Name
terrestrial	Tara Strall
conifer	
monocot	
chlorophyll	
anther	
pollinate	
meristem	

5. Find *two* common two-letter words from each of the listed terms. You may use a letter only once. The first one is done for you.

Term	Two-Letter Words	
photosynthesis	to	he
pollination		
conifers		
internode		
stomata		
venation		

6. Write the names of plants that sound like the listed nonsense words. The first one is done for you.

Nonsense Words	Plant
Zirhn	fern
riversnorts	
forcepale	
moonapsur	
spysleds	
rinkmoes	
swaypull	

PLANT LIFE (continued)

18–4 Create-A-Comment

Students can relax and let their imaginations roam. Creating comments for sketch figures is a fun way to think about science.

Give students a copy of an illustration and have them decide what comment or statement to write in the balloon above the two figures.

1.

2.

PLANT LIFE (continued)

18–5 Riddle Bits

Solving riddles can be frustrating for some students. However, active minds produce interesting results. Frustrating or not, let students tackle the 12 riddles and see how successful they can be.

1. What vegetable comes with its own dinnerware?
2. What vegetable has an insect as part of its name?
3. If a flower fails to get pollinated, what does it send out?
4. What plant contains a valuable metal in its name?
5. What bryophyte has an organ and the unfermented infusion of malt in its name?
6. What do the books titled *My Friend Flicka* and *Black Beauty* have in common with plants?
7. What tracheophyte's name hints at plant abuse?
8. Why do large fern leaves stay close together?
9. Mr. Leaf Rhizoid, attorney, grew ferns—lots and lots of ferns. People accused him of taking his hobby to work with him. How could this be?
10. What kind of country would we have if plants reproduced at the same time?
11. What is a conifer?
12. Why are some trees nosier than others?

18–6 What Goes Where?

Have students unscramble the letters (if necessary), decipher the words and phrases, and identify the symbols. Then they can organize everything to reveal the mystery term.

1. lird + s + 10.
2. [flag] + 5th month.
3. "q" + el + tick.
4. "t" + "p" + elo.
5. "oo" + ti + one-third of *freeze*.
6. $\dfrac{\text{plant}}{\text{e}}$.

Life Science

PLANT LIFE (continued)

18-7 Creative Potpourri

This section offers five miniactivities of miscellaneous nature for students to try. The activities range from simple problems to interpreting pun statements.

1. The letters needed to spell the two mystery vegetables are crowded together. See if you can separate the letters and reveal the answers.

   Vegetable 1   Vegetable 2

2. Name five plants whose names relate in some way to an animal.

   Example: *peti*ole - pet

3. How many seeds are present in the diagram?

4. Find a description to match each of the rhyming words. For example, an *epidermis thermos* is what prevents a plant from freezing. Be playful with words. Have fun and let your imagination be your guide.

spore pour	thistle pistil
lone cone	fruit root
pollen fall'en	moss toss
dud bud	phloem dome

5. A pun plays upon words similar in sound but having a different meaning. Read the following pun statements and record what you think they mean.

   a. Did you hear about the desperate farmer who needed a loam?

   b. And then there was the anther who became stigmatized for trying to keep in style.

Section 19

ANIMALS WITHOUT BACKBONES

19–1 What Does the Sketch Represent?

The same sketch may bring 30 or more different interpretations. A creative, flexible thinker may see several ways to describe a scene or situation.

Have students select a sketch and examine it carefully. Then ask them to record what they think the illustration represents on the line below the drawing.

1.

2.

3.

4.

5.

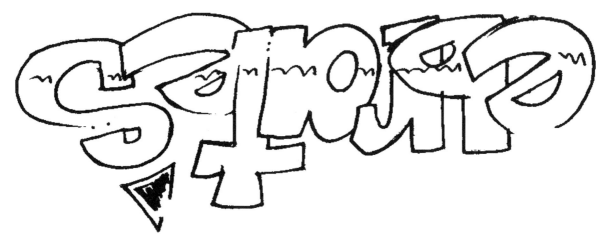

_____

6.

An Expression

7.

_____

8.

ANIMALS WITHOUT BACKBONES (continued)

19–2 Can You Solve the Problem?

Most students find miniproblems a friendly challenge and worth solving. A helpful hint along the way keeps the frustration level under control.

1. Cnidarians are water-dwelling animals. Two examples are the sea anemone (salt water) and hydra (fresh water).

 The scattered letters in the box make up the names of two marine (saltwater) cnidarians. Use each letter once to write the names in the spaces below the box.

```
         e       r
          h     y
       a         o   j
    s   i   c   l   f
                       l  s  l
```

_____   _____

2. Use the letters in the word COELENTERATA to write ten words starting with the letter *t*. You may use a letter more than once.

3. The grasshopper is a classic insect. It has three distinct body parts, six legs and wings. Write the names of ten different insects using the letters in GRASSHOPPER to complete the spellings. Write the names of insects across the letters in GRASSHOPPER (see the following example).

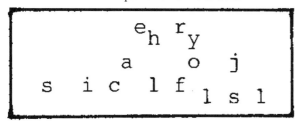

4. Animals in the phylum Mollusca are missing something. What is it? To find out, you must answer each of the three questions with a number. After you answer a question, darken all the spaces in the puzzle that contain the number. If you answer the three questions correctly, the darkened spaces will reveal what the mollusks are missing.

 Questions

 a. How many legs do insects have?
 b. How many body sections does a spider have?
 c. How many pairs of legs does a centipede have attached to each body segment?

Life Science

ANIMALS WITHOUT BACKBONES (continued)

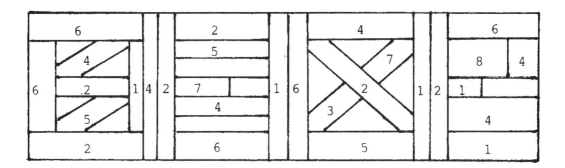

5. There is a *large* vertebrate animal trying to disguise itself as an invertebrate. Its name is hidden in the phylum ECHINODERMATA (spiny-skinned sea animals). See if you can find it. *Hint:* It's not a cat.

6. Starfish regenerate or replace lost parts. Write the word STARFISH in the spaces. Begin with the letter *s* in the far left square.

   How many spellings did you get?

   What should be the final starfish count? Why do you think so?

7. An endoskeleton is an internal or inside skeleton. Echinoderms have endoskeletons. An exoskeleton is an external or outside skeleton. Arthropods possess an exoskeleton.

Show, using a diagram, how an *endoskeleton* might occur exactly in the middle of an exoskeleton.

19–3 What's in a Name?

This section offers seven miniactivities using a plethora of words, terms, and names. Imaginative thinking is a key factor for success.

1. Creative thinkers arise. See how many first and last names you can create out of the eight *arthropod* structures. The first one is done for you.

   Example: antennae - Ann Tinnay

abdomen	cornea
maxillae	ommatidium
carapace	tympanum
rostrum	halteres

ANIMALS WITHOUT BACKBONES (continued)

2. Tod and Mat Hydra are brothers. They share something unusual about their first names. If you can figure it out, you're a great detective. Use the clues to help you solve the mystery.

    Clues
    - Everything you need to know is in their names.
    - Hydra *structures* are most revealing.

3. The phylum Annelida includes segmented worms. An earthworm is an example of an annelid. Five girls' names are hidden in the word *Annelida*. Can you find them? You can't move letters around, but you can use a letter more than once.

4. If invertebrates could speak English, which of them might say the following?

Expression or Statement	*Invertebrate*
a. "Let's go for a spin."	_____
b. "Wait. I'll get my wrap."	_____
c. "Don't try to wiggle out of it."	_____
d. "It'll do in a pinch."	_____
e. "You little blood sucker."	_____
f. "I'm absorbed in the story."	_____
g. "She rubs me the wrong way."	_____
h. "Hop to it!"	_____

5. How many invertebrates can you name that have the "ca" letter combination in their names?

6. What *two* insects have the names of members of the Order Acarina in their names?

7. Members of the Class Insecta (insects) are the most abundant of all land animals. Write the name of an object or thing found *in* or *around* the house that rhymes with each of the listed insects. For example, moths eat clothing, thus *moth cloth*.

    bee   cricket
    ant   bug
    flea   fly

Life Science

ANIMALS WITHOUT BACKBONES (continued)

19–4 Create-A-Comment

It's great fun creating humorous statements or comments to match the action of an illustration. The following two drawings need help: they're begging to be completed. Therefore, select an illustration. Examine it carefully and write a statement or comment in the balloon above the figures in the sketch.

1.

2.

Life Science

ANIMALS WITHOUT BACKBONES (continued)

19–5 Riddle Bits

Anyone for a riddle? If so, have students take a crack at any of the 15 riddles waiting to be tackled.

1. What insect is a carrier of rickets?
2. There are at least three times when a fish is not a fish. What are these times?
3. What insect has its name in part of its body structure?
4. What insect is the most well read?
5. What invertebrate is made up of about 83 percent nails?
6. What worm do scientists believe evolved from primates?
7. What winged insects are known as "fibbers in flight?"
8. What ocean-dwelling crustacean came from the country?
9. What is the difference between incomplete and complete metamorphosis?
10. What are pheromones?
11. What do you call it when a bee stings somebody?
12. What should you call a parasitic trematode worm that "lucks out" at the right time?
13. How is a wristwatch like a female Acarina?
14. How is a wet, soggy carpet like the letter *g* in sponge?
15. What two crustaceans can be found in a medicine cabinet?

19–6 What Goes Where?

Creativity comes to life as students scramble to unscramble a collection of symbols, words, and numbers. Have them identify the symbols, decipher the words and phrases, and uncover the hidden items.

1. in + prefix for three + ch + a.
2. [hammer symbol] + s.
   [symbol]
3. second word   second word
   [face] + ru   "c" (two words).
4. 8's + "sole" + opposite of exit.
5. (Fe + O = ?) + c + sna + playing card, single spot.
6. su + "not" + [eye symbol] + l.

ANIMALS WITHOUT BACKBONES (continued)

19-7 Creative Potpourri

Students can try their hands at interpreting puns, solving minipuzzles, and writing silly, strange phrases and sentences.

1. Use the following terms as silly, sound-alike words to construct weird sentences.

    Example: "Will I go? Nautilus (Not unless) you do."

    porifera   nematode   katydid   trichina   mussel

2. A pun plays upon words similar in sound, but having a different meaning. Read the following pun statements and record what you think they mean.
    a. Then there was the porifera who was removed from the colony. He was becoming too much of a sponge.
    b. The best music group at the Crawdad Festival was *Kara Pace and the Swimmerets*.

3. Find the "lucky seven" invertebrate members who have been selected to participate in this activity. Two clues are provided to help you identify *each* member. Write the name of each organism in the space to the right of the clues.

Clues	*Invertebrate Members*
a. sea, "flower"	_____
b. pores, hollow	_____
c. roundworm, parasite	_____
d. parasite, segmented	_____
e. oyster, pearl	_____
f. echinoderms, spines	_____
g. coleoptera, insects	_____

    What do you notice about the "lucky seven"?

4. Use the clues to identify the three invertebrate sea creatures. Then write the letters to their names in the numbered spaces in the diagram. Finally, answer the question: What can you say about these sea critters? (*Hint:* It's an expression; *Think* "circle.")

ANIMALS WITHOUT BACKBONES (continued)

Clues	*Diagram Numbers*
pincers, some are hermits	1 through 4
bivalve, ridged shell	5 through 11
mollusk, spiral shell	12 through 16

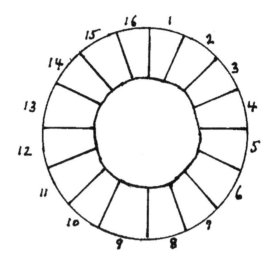

5. Look at the box carefully. How many times will a *spider* appear? Count the letters, but use each letter only once.

a	d	i	s	r	s	e
i	p	h	i	p	r	s
n	c	d	p	e	a	p
i	p	i	s	e	r	i
s	d	d	e	r	d	r

Section 20

ANIMALS WITH BACKBONES

20-1 What Does the Sketch Represent?

One way to have fun in science is to create strange, humorous titles or descriptions for sketches. Students can relax and take a tongue-in-cheek adventure with one, two, or more of the following illustrations.

Ask students to study a sketch. Then have them write what they think the illustration represents on the line below the sketch.

1.

278

2.

3.

4.

5.

6.

ANIMALS WITH BACKBONES (continued)

20–2 Can You Solve the Problem?

This section offers students eight miniproblems to solve. A pound of patience mixed with five pounds of determination should provide enough ammunition to be a winner.

1. Solve the great fish mystery. Read the five clues describing the fish's external features. If you unscramble the circled letters, you will reveal the answer.

    Clues

    - The snout has two doub(l)e olfactor(y) sacs ("nostrils").
    - There ar(e) two se(p)arate dorsa(l) fins.
    - The b(o)dy is spindle-shaped, higher than (w)ide.
    - A larg(e) mout(h) with fine teeth.
    - There are dermal s(c)ales on the t(r)unk and tail.

2. Show how *all* amphibians have two eyes.
3. Show how the Subclass Caudata (salamanders) dominates man.
4. Show a feline that *doesn't* resemble a domestic cat or lion or tiger or leopard.
5. Show how to create two hyenas in a box using eight letters.
6. Show how a bird's body is divided into a head, neck, trunk, and tail.
7. Show how only you can reveal the difference between a true toad and a tree toad.
8. A mammal, in order to stay in one piece, should be concerned with the "Three M's." What are the "Three M's?"

20–3 What's in a Name?

The seven activities in this unit center on the relationship among various words, terms, and names. These miniactivities keep the thinking process alive and well.

1. Write the name of a bird that best matches each comment. Remember, it's okay to rev up the imagination and push the humor button.

    Example: (Comment) "He's a member of the flock." (name of bird - goose)

Comment	Name of Bird
"Stick 'em up."	_____
"Lower your head."	_____
"What an easy mark."	_____
"What a crazy person."	_____

ANIMALS WITH BACKBONES (continued)

2. Fill in the empty spaces in the chart with rhyming names of land or water vertebrates.

LAND	WATER
	hake
snail	
lark	
	seal
woodchuck	
	swan
	frog
mole	

3. Find an organism's name hidden in each of the following vertebrate names.

manatee	albatross
elephant	cassowaries
stickleback	fowl
chickadees	boar
pigeon	sphenodon

4. In each of the listed vertebrates, two-, three-, or four-letter combinations form part of the name of a human structure. Write the name of the structure in the space to the right of the vertebrate. *Underline* the letter combinations in the human structures. The first one is done for you.

Vertebrate	*Human Structure*
crane	*cran*ium
manatee	_____
cod	_____
fowl	_____
deer	_____
chimp	_____
armadillo	_____
chipmunk	_____
bear	_____
skunk	_____

ANIMALS WITH BACKBONES (continued)

5. List eight vertebrate mammals that have the *ea* letter combination in their names.
6. List eight vertebrate mammals that have the same two vowels next to each other in their names.
7. Write sentences using each word combination. Use both words in the form they appear.

    ape, grape   whale, sail
    lemur, femur   owl, towel
    shark, dark   hawk, rock

ANIMALS WITH BACKBONES (continued)

20–4 Create-A-Comment

Creating cartoon humor can get the neurons dancing. The following two activities allow students to be unique in their thinking and develop clever comments and statements to support the section theme.

Provide students with a copy of a drawing. Ask them to write a response—comment or statement—in the balloon above the two figures in the illustration.

1.

2.

ANIMALS WITH BACKBONES (continued)

20-5 Riddle Bits

Attempting to solve riddles feeds an active brain and gives it room to dig deep for unusual, albeit bizarre, responses. Creative students seek revenge. They often test the teacher's ability to solve "student-made" riddles.

1. What rodent dominates animals with backbones?
2. Where do "upials" come from?
3. What mammal suffers from constant depression?
4. What giant mammal needs an ant to keep itself from breaking apart?
5. What mammal runs away with its mate?
6. What reptile depends on an ocean fish to keep it intact?
7. What reptile is always linked with a civil or private wrongdoing?
8. What group of animals suffer a painful youth?
9. What would you get if you crossed a dollar bill with a kangaroo?
10. How is an infant like a mother rattlesnake?
11. What reptile has a fluid material used for writing as part of its name?
12. What mammals are besieged with rage?

20-6 What Goes Where?

Something different. A change in direction. Give students the following six activities, each a scrambled collection of words, symbols, and an occasional number. Have them unscramble the mixture and reveal the mystery term.

1. a + a + 2 + rat.
2. 🗝 + ru + t.
3. "mmm + r + mmm" + set + "oh."
4. ara + teek + p.
5. th + opposite of off + π.
6. 💡 + opposite of him.

ANIMALS WITHOUT BACKBONES (continued)

20–7 Creative Potpourri

The final section offers students five activities ranging from pun statements to a miniproblem. These activities make excellent closers for the last few minutes of a class period.

1. A pun plays upon words similar in sound but having a different meaning. Read the following pun statements and record what you think they mean.

    a. And then there was the neurotic bear who developed a TIC.

    b. A migrating bird will travel far if given enough latitude.

2. Create cartoon illustrations for these captions:

    "Sorry, Frank. You'll have to fin for yourself."

    "You're just an old lump."

    "He's a snake in the grass."

    "It's raining cats and dogs."

    "Boy, there goes a real quack."

3. Create slogans or expressions related to vertebrates.

    Example: Chickens develop a fowl attitude.

4. Create silly girl/boy first and last names out of these vertebrates:

tadpole	coelacanth
salamander	lizard
amphibian	alligator
elephant	giraffe

5. There is a mammal hiding in the word MAMMAL. Can you find it? *Hint:* You may move letters around and use one letter twice.

Section 21

REPRODUCTION AND DEVELOPMENT

21-1 What Does the Sketch Represent?

Finding descriptions for sketches will keep the creative pump house busy churning out an idea or two.

Ask students to examine a sketch. Then have them write what they think the drawing represents on the line below each illustration.

1.

291

2.

3.

4.

5.

6.

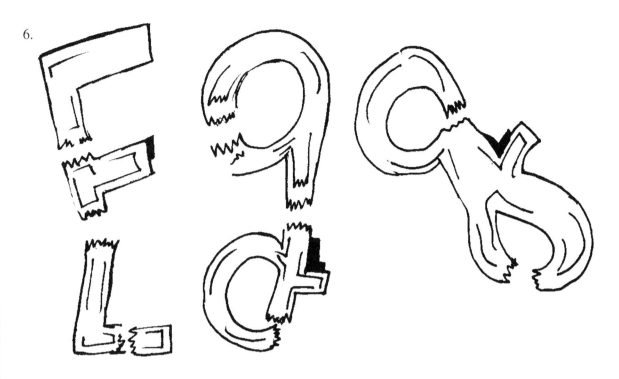

7.

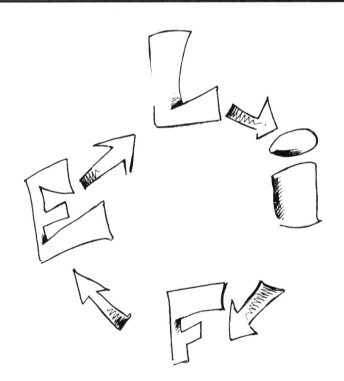

8.

Life Science 297

REPRODUCTION AND DEVELOPMENT (continued)

21–2 Can You Solve the Problem?

Have students get in a creative mood by tackling the following six miniproblems.

1. Use all the letters in the box to write the names of four structures related to the female reproductive system.

c	v	r	c	o	i	t
i	y	u	o	v	u	r
v	s	x	e	d	a	t
		r	e	u		

2. Use all the letters in the box to write the names of four structures related to the male reproductive system.

   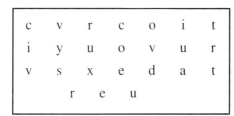

3. Use parts of each of the three words to make a word that describes a human embryo from approximately five to ten weeks *after* conception.

   detach   budding   envelope

4. Reproduction is the process by which a plant or animal gives rise to another of its kind. List 12 plants or animals that reproduce members of its kind. Each organism must begin with the letter listed to the left of the space.

REPRODUCTION AND DEVELOPMENT (continued)

R _____
E _____
P _____
R _____
O _____
D _____
U _____
C _____
T _____
I _____
O _____
N _____

5. Unscramble the five terms. Then combine the middle letters of each term to form a word that will answer this question:

   What is the fluid that delivers oxygen and nutrients to the developing embryo?

   romeybs _____   rmsucot _____
   lbamicuil _____   bdolo _____
   tudvoci _____

6. You have three words: embryo, fetus, and infant. Some of the letters that make up these three words will produce the answer to this question:

   If everything goes right, what will the growing child eventually do?

   Hint: Use *three* letters from *embryo*, *one* letter from *fetus*, and *two* letters from *infant*.

21–3 What's in a Name?

Students can fine-tune their cerebral machines by manipulating words to form names or by using their imaginations to create names. Here are five activities to whet their appetites.

1. Make up rhyming first names of boys and/or girls from the listed terms. You may abbreviate names; however, do not use nicknames. The first one is done for you.

   Example: fetus - Cletus Fetus

   egg   vessel   age
   cell   boy   gland
   male   girl

Life Science 299

REPRODUCTION AND DEVELOPMENT (continued)

2. Boy and girl names appear in column A. Find a term related to reproduction and development that has the name as part of the term.

    Example: Art - artery

Column A	Reproduction and Development Term
Bert	_____
Nancy	_____
Mary	_____
Epi	_____
Liza	_____
Liz	_____

3. Use 22 letters from the eight names to identify three stages of human development. You may use a letter *only once* from the same name.

    Names
    Thomas
    Danielle
    Georgette
    Nancy
    Frederick
    Antonio
    Helen
    Donna

    Stages of Development

    a. _____ (seven letters)

    b. _____ (nine letters)

    c. _____ _____ (six letters)
           (two words)

4. Meet Liz Andrews, a hypothetical person for our activity. Use the letters in her name as the first letters to identify ten terms related to reproduction and development. The first one is done for you.

    L life cycle
    I _____
    Z _____

    A _____
    N _____
    D _____
    R _____
    E _____
    W _____
    S _____

REPRODUCTION AND DEVELOPMENT (continued)

5. Find a boy or girl's name (nicknames are okay) hidden in each of the terms.

Terms	Name(s)
a. development	1.
b. progesterone	2.
c. pregnancy	3.
d. oviduct	4.
e. umbilical cord	5.
f. reproduce	6.
g. fallopian tube	7.

Life Science   301

REPRODUCTION AND DEVELOPMENT (continued)

21–4 Create-A-Comment

The following two activities offer students an opportunity to see the humor in a cartoonlike sketch.

Give students a copy of a drawing. Have them examine the sketch. Then ask them to write a comment or statement in the balloon next to the figures in the illustration.

1.

2.

Life Science 303

REPRODUCTION AND DEVELOPMENT (continued)

21–5 Riddle Bits

Do your students have the stamina to tackle challenging riddles? Here's a chance to find out. Give them the following riddles to bounce around.

1. If *di* and *et* were the same, what would you have?
2. What do you call a newly hatched ant?
3. How can a 100-pound mother have a 40-pound offspring?
4. How can a 100-pound mother have a 40-pound baby?
5. Bill Brothers had two sisters and three additional family members. How many brothers were in Bill's family?
6. Mrs. Wynn had just given birth at the county hospital. She said, "My husband Tom and I have just hit the jackpot—twice." What do you think she meant?
7. Thirteen (13) babies were born in St. Fictitious Hospital. The only clue to birth records is as follows:

   ```
   Allen,  Betty
   Amory,  Donna
   Boyce,  Sb
   Gerard, Lisa
   ```

   How many babies were boys?
8. There was a man holding a sign at a protest rally. The sign looked like this:

   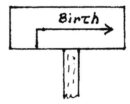

   What do you think the sign meant?
9. What do you call the container or capsule that carries the egg?
10. What six-letter word makes up PARENT?
11. What would you get if you combined two "ults"?

REPRODUCTION AND DEVELOPMENT (continued)

21–6 What Goes Where?

Have students use their creative detective prowess to unscramble the symbols, phrases, letters, or numbers to reveal the correct terms.

1. (fe + O = ?) + you + 2.
2. Los Angeles (abbr.) + a + What a penny's worth + p.
3. burr + e + m + o.
4. s + u + 🦶 🦶
5. chemical symbol, Ne + ma (spelled backward).
6. in + chemical symbol for silver + g (first word); ssec (backward) + opposite of amateur (second word).

21–7 Creative Potpourri

These five activities will stir up the neurons and keep students mentally alert.

1. Examine the different shapes joined together in the box. A nine-letter word can be created by forming letters from the design. What is the word? *Hint:* It means to bring forward.

2. Find one term related to reproduction and development for *each* double letter combination that makes up the word in column A. The first one is done for you.

Column A			Terms	
a. rice	ri	ov<u>a</u>ries	ce	cerv<u>i</u>x
b. bean	be	_____	an	_____
c. stem	st	_____	em	_____
d. root	ro	_____	ot	_____
e. dirt	di	_____	rt	_____
f. pith	pi	_____	th	_____

Life Science

REPRODUCTION AND DEVELOPMENT (continued)

3. Use the letters in the term to identify the mystery name. The description will provide a clue to the correct answer. You may use a term, word, or letter more than once.

Term	Description	Mystery Name
a. urethra	a western state	_____
b. uterus	public road	_____
c. puberty	pleasing, neatly arranged	_____
d. epididymis	move quickly; velocity	_____
e. hormone	space object	_____
f. estrogen	a list of names	_____

4. Show, using diagrams, what you would get by crossing the following organisms: horse and coral, snail and snake, and gorilla and swine.

5. Find *four* words from the letters in FERTILIZATION that rhyme with *tile*. You may use letters more than once.

Section 22

Genetics and Change

22–1 What Does the Sketch Represent?

Describing an illustration based on a selected theme takes imagination and a willingness to accept a bizarre thought or two.

Ask students to examine a sketch and write what they think the illustration represents on the line below the drawing.

1.

2.

3.

4.

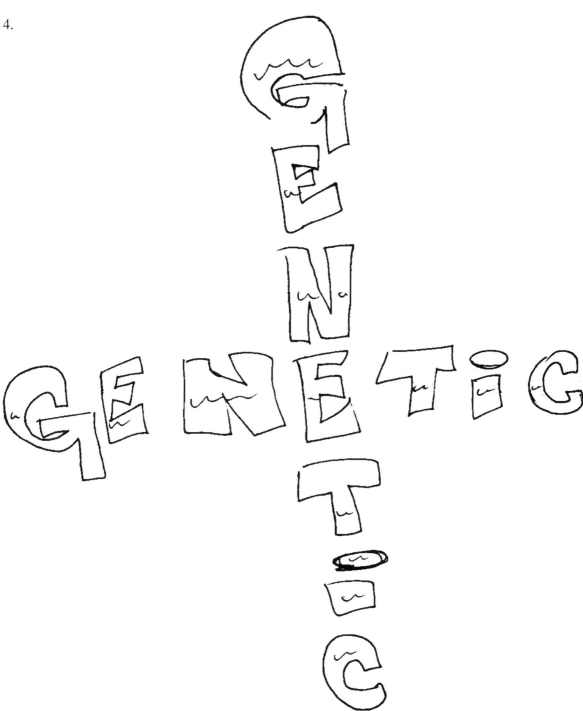

5.

6.

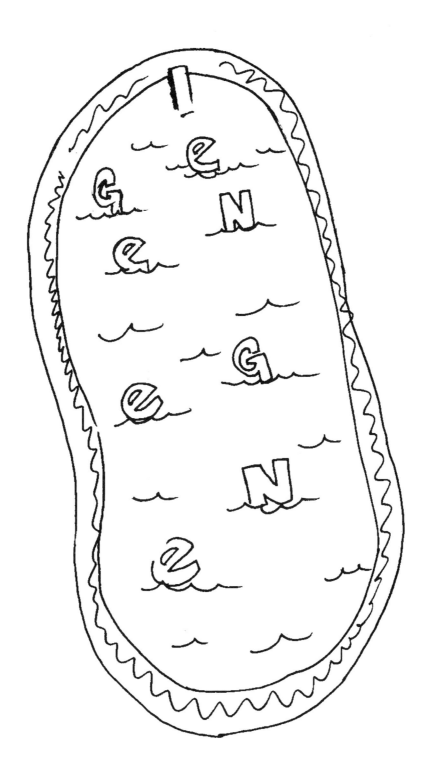

7.

8.

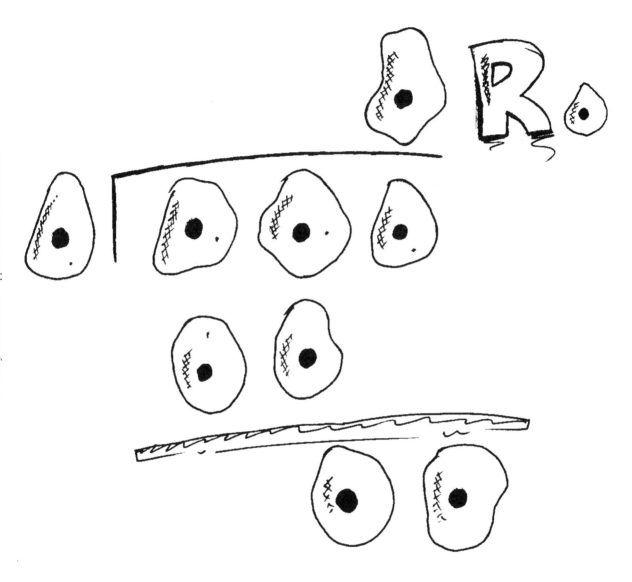

GENETICS AND CHANGE (continued)

22–2 Can You Solve the Problem?

Problems are fun to solve, especially for those students who are anxious to stretch their neurons to the limit. The five miniproblems will help warm up the cerebral machine.

1. Bill didn't know what the word GAMETE meant. His teacher, Mr. Spencer, offered Bill a challenge. He would treat Bill to lunch if he could find the definition *and* use some of the letters that spell GAMETE to write the name of an example of a gamete. However, there was a catch: Bill had a one-minute time limit.

See if you can meet the challenge. *Hint:* You may use the same letter more than once.

2. How many DNA's can you make from the letters scattered around the box?

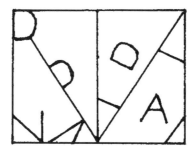

3. What do you think these figures are? *Hint:* Unscramble the letters.

4. How can you show INCOMPLETE DOMINANCE by using the letters M N D and E?
5. A person runs an ad to sell his dog. He wants to save money on the cost of the ad. The charge: fifty cents per letter. So he runs this ad:

   BEDER dog for sale. $30. Call 473–8260.

   What kind of dog is for sale?

   How much money does the person save?

Life Science

GENETICS AND CHANGE (continued)

22-3 What's in a Name?

Have students select any of the six activities to test their ability to rearrange letters and discover the correct answers.

1. Look at the word GAMETE. The letters may be rearranged to produce two words that describe nondangerous animals hunted for food or sport. A letter may be used more than once.
2. Find the name of a nitrogen base that can be formed from the letters in DEOXYRIBO-NUCLEIC ACID. A letter may be used more than once.
3. Use the letters in the box to spell the first and last name of the Austrian man who studied inherited traits. Use 12 of the 16 letters.

If you unscramble the remaining 4 letters, you'll discover the position or title held by the man.

| N | M | E | D | G | N | R | L | G | M | E | O | R | E | K | O |

4. Rearrange the letters in each term to create a boy or girl's first and last name. Use your imagination and have some fun. The first one is done for you. You must use all the letters. However, you may use a letter only once.

   Example: Meg Eta (gamete)

   phenotype   adenine
   mutation   cloning
   guanine   autosome

5. The left-hand column provides a list of two-word descriptions. The matching terms in the right-hand column consist of scrambled letters. Unscramble the letters and write the name of the matching term in the space to the left of the number.

   Two-Word Descriptions   *Scrambled Letters*

   _____ a. weaker genes   a. tdnomnia
   _____ b. trait carriers   b. enseg
   _____ c. trait chart   c. iepdereg
   _____ d. heredity study   d. sgcenite
   _____ e. stronger genes   e. vescreise

6. Remove the first two letters of each listed term. Then rearrange the remaining letters to produce a recognizable word.

   hybrid   chance
   clone (don't use *one*)   inherit
   gamete   animal

GENETICS AND CHANGE (continued)

22–4 Create-A Comment

Have students create their own comments related to the cartoon theme. Provide them with a copy of a drawing. Ask them to examine the drawing and write a comment or statement in the balloon next to the figure in the illustration.

1.

2.

GENETICS AND CHANGE (continued)

22–5 Riddle Bits

Offer students these riddles to juggle about. The riddles provide a wonderful exercise to warm up the creative spirit.

1. What kind of ability does a person have who solves problems in a "hit or miss" fashion?
2. What do you call an event where "gams" gather together?
3. Why do geneticists usually lose at poker?
4. Where do geneticists go shopping?
5. What would you get if you crossed a bat with a rat?
6. What would you get from a typewriter with only these keys—A, B, and O?
7. You see three identical masked men on white horses riding across the plains. What are you observing?
8. You see a 300-pound parrot sitting on the shoulder of a 500-pound person. What do you think is the parrot's name?
9. This sign suggests self-pollination in plants. What does it mean?

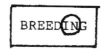

10. What would you get if you crossed a slime mold with a bug?
11. What would you get if you crossed "Le Fly" with "Le Frog"?
12. What would you get if you crossed a stork with a shrew?

22–6 What Goes Where?

Have students identify the symbols and find the missing letters. Then if they arrange everything correctly, they'll uncover the mystery terms.

1. ![fish] + 50 cents to park + l.
2. To correct for publication + y + she (possessive case).
3. ![coils] + opposite of on.
4. 8's + trunk (-unk).

Life Science

GENETICS AND CHANGE (continued)

5. l's + chlorine (chemical symbol).

6. l + to repair + e.

22–7 Creative Potpourri

Offer students a mixture of puns, challenging definitions, bizarre sentences, and terms to keep their creative wheels turning.

1. Ready for a pun? A pun plays upon words similar in sound, but having a different meaning. Read the following pun statements and record what you think each one means.

    a. How about the evolutionist who looked so much like his father that people said he was a "chimp" off the old block.

    b. It's true Mendel remained single. He was afraid of becoming dominated.

    c. Did you hear about the geneticist who was fired for "cloning around"?

2. Make up silly definitions for the following terms:

    Example: hybrid - a much taller organism than a "lobrid"

genetics	chromosome
recessive	genetic code
self-pollinate	pedigree
DNA	gene splicing
polyploidy	cloning

3. Find the genetic term hidden in each sentence. The first one is done for you.

    a. Jean wanted to do something ex*tra. It* was for Mr. Jones' math class. (. . . ex*tra. It* . . . = *trait*)

    b. Michelle asked Mary how an elephant manages to get around. Mary replied, "BIG ENERGY."

    c. Mark said Kim would succeed in life. "She has it in her. It's a talent."

    d. Some men delegate responsibility; others try to do everything themselves.

    e. Bob wanted Janice to play in the game. "Tetherball," he said, "is fun for everyone."

4. Make a complete sentence out of each word. First, however, you must divide the word into two separate words. Then use both words in the sentence and *underline* them.

GENETICS AND CHANGE (continued)

Example: GENETICS - <u>Gene</u> has developed several <u>tics</u>.

factor	dominate
offspring	dominant
autosome	genotype

5. List ten genetic terms that have the "in" letter combination.

Section 23

Human Biology

23–1 What Does the Sketch Represent?

Ask students to examine a sketch or two and write what they think the sketch represents on the line below the drawing. Their responses should match the section theme.

Flexible thinkers do well creating descriptions for illustrations. There are no right or wrong answers. So every student should experience success.

1.

321

2.

3.

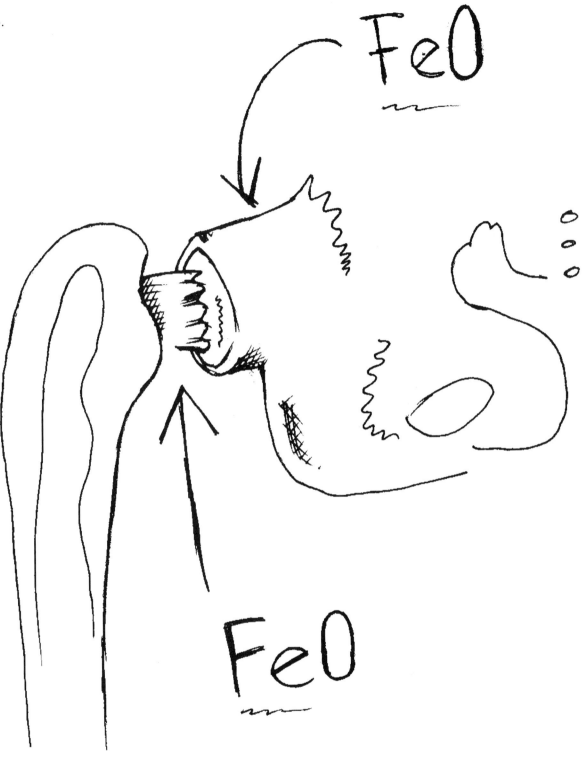

4.

5.

6.

SYSTEM

7.

8.

Life Science

HUMAN BIOLOGY (continued)

23-2 Can You Solve the Problem?

Miniproblems are fun to solve. They take 2 to 5 minutes to complete and fully engage the thinking process. And perhaps best of all, they serve as stimulators for classroom discussion.

1. Unscramble and combine the chemical symbol of each element to form a word that relates in some way to the *human body*.

    Example: Carbon, neon and potassium-C, Ne and K = *NeCK*.

    a. oxygen, nitrogen, and selenium
    b. indium, sulfur, and potassium
    c. neon, boron, and oxygen
    d. phosphorus and lithium
    e. selenium and nobelium
    f. astatine and helium
    g. iron and lithium

2. There are five letters pressed together in the box. See if you can find each letter. Then put them together to spell a word. What word? Here's a hint: MAN, OH MAN!

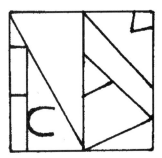

3. Can you find two red blood cells hidden in the letters?

```
o   s   l   l   c   l
c   l   o   t   e   d
b   o   d   e   c   e
s   d   e   r   d   o
b   l   w   o   l   r
```

HUMAN BIOLOGY (continued)

4. Write a word that rhymes with each of the listed body structures in the space to the right of the word. Use the hints to find the correct answers.

Body Structures	Hints	Rhyming Word
a. kidney	Australian City	_____
b. blood	Noah's ark	_____
c. foot	Anchors a plant	_____
d. muscle	Rush hurriedly	_____
e. brain	Long-necked bird	_____
f. digest	Take into body	_____
g. gland	Smooth, mild	_____
h. heart	Sharp, sour taste	_____
i. vein	Discoloring	_____
j. knee	Run away	_____

5. List eight terms related to the human body that have an "or" or "ro" letter combination.

6. The brain assists you in many ways. Write the fourth rhyming word in the space to the right that indicates what the brain helps you do.

   a. ink, drink, sink, and _____
   b. deal, meal, teal, and _____
   c. fear, steer, gear, and _____
   d. free, tree, knee, and _____
   e. fell, tell, bell, and _____
   f. groove, behoove, Louvre, and _____
   g. seethe, heave, cleave, and _____
   h. stain, rain, train, and _____

7. The human skeleton consists of 206 bones. Use the list of hints to locate the mystery bone.

Hints	Mystery Bone
a. Holds Presley's name	a.
b. The first part of the name is a "lie about something unimportant."	b.
c. Three letters suggest a covering for the head.	c.

Life Science   331

HUMAN BIOLOGY (continued)

    d. A cone-bearing tree makes
       up most of its name.   d.

    e. A dark, oily hydrocarbon makes
       up half its name.   e.

    f. Its name includes an adult male
      human being.   f.

    g. Half its name indicates an alcoholic
       liquor distilled from fermented
       molasses or cane juice.   g.

23–3 What's in a Name?

Creating names or finding names within terms requires a sufficient cerebral effort. The following activities make excellent closers (the last 5 minutes of the class period) as well as starters (the first 5 minutes of the class period).

1. Write each of the seven terms backward. Then find a boy's or girl's name or nickname in the backward spelling. Write the name in the far right column.

   Term   *Term Spelled Backward*   *Name or Nickname*

   a. patella

   b. clavicle

   c. spinal cord

   d. dermis

   e. vessel

   f. parathyroid

   g. adrenalin

2. There are boy's and girl's names and nicknames hidden within each of the listed names. Find the number of names indicated in parentheses. Then write the names in the spaces.

   Descartes (3)

   Vesalius (2)

HUMAN BIOLOGY (continued)

Galileo	(3)	_____	_____	_____
Crick	(1)	_____		
Mendel	(1)	_____		
Redi	(3)	_____	_____	_____
Beaumont	(1)	_____		
Schwann	(1)	_____		

3. List five body structures or conditions that have the word *art* in their names.
4. Unscramble the terms that name five bones located from the shoulder to the knee. Then write each term in the space next to the rhyming word.

Scrambled Word	*Rhyming Word*	*Bone Name*
a. esnpi	farces	_____
b. tlaplea	typhoid	_____
c. reumf	umbrella	_____
d. srtuas	lemur	_____
e. dipoxhi	sublime	_____

5. List 12 body structures that have *three* letters in their names (singular).
6. List 12 body structures that have *four* letters in their names (singular).
7. List seven bones that have the letter "p" in their names.

Life Science   333

HUMAN BIOLOGY (continued)

23–4 Create-A-Comment

These two activities allow students to create humorous comments to match the illustrations.

Give students a copy of a drawing. Ask them to examine the sketch and write a comment in the bubble next to the figure.

1.

© 1993 by The Center for Applied Research in Education

2.

Life Science

HUMAN BIOLOGY (continued)

23-5 Riddle Bits

Riddles get things going quickly. They ignite the thinking machine and produce interesting, albeit weird, responses.

1. What is humorous about the humerus?
2. What part of cartilage is esthetically pleasing and meaningful?
3. How are pores like rocks containing valuable metals?
4. Why is the law firm of Ischium, Ilium, and Sacrum so popular?
5. If muscles could speak, which ones would say "I'll do it! I'll do it!"?
6. Where did many students digest the greatest amount of information?
7. If a person received an award for not brushing his teeth, what kind of a reward would it be?
8. If someone blamed the moon for a broken romance, what may have gone wrong?
9. What "plant" has its own breathing system?
10. What is the most recently discovered respiratory disease?
11. What alcoholic liquor remains absorbed in the brain?
12. What *number one* structures appear in all parts of the human skeleton?

23-6 What Goes Where?

Give students the following collection of symbols, words, and numbers to examine. Ask them to identify the symbols and decipher the words and phrases. Finally, have them rearrange the items to reveal the correct term.

1. 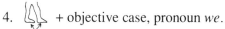 + br.
2. um + 🌳 + a.
3. another word for sick + constellation, ram + 👤
4. 👃 + objective case, pronoun *we*.
5. live (–e) + a + salad (–lad).
6. s + another name for mom + play (–y).
7. chemical symbol (Sb) + a + red (–d).

23-7 Creative Potpourri

The following activities are a mixture of puns, silly definitions, weird sentences, and the dissection of a few words. Students should enjoy the challenge.

HUMAN BIOLOGY (continued)

1. Try your skill with a pun or two. A pun plays upon words similar in sound, but having a different meaning.

Read the following pun statements and record what you think they mean.

   a. It has been said that a bone break in the upper arm is really humerus.
   b. The pituitary is often referred to as the Master Gland. Well, it's only a hypophysis.
   c. An Adam's apple is Eve's revenge for the ribbing she got.

2. The STERNOCLEIDOMASTOID muscle controls head and neck movement. Use the letters in the word to write the names of *two* well-known objects in space. You may move letters around, but you can use a letter only once unless the word has two or more of the same letters.

   Answers: (1) _____ (2) _____

Now that you've found the two answers, six letters remain. If you unscramble five of the six letters, you'll have the name of a phase or state of matter.

   Answer: _____

Finally, you wind up with one letter. What is the letter? The remaining letter is _____

3. ENDOPLASMIC RETICULUM is a network of tubes within the cytoplasm of a cell. Use the letters in the words to write the names of eight bones. You may use the same letter more than once. *Challenge:* Write the name of an upper body bone from the letters in ENDOPLASMIC RETICULUM. You may use a letter only *once*.

4. See how good you are at creating silly "scientific" definitions for human biology terms. The first two are done for you.

   Examples: ilium - an unhealthy "ium"
   extensor - a retired "tensor"

enzymes	intestine
peristalsis	ventricle
artery	bronchus
zygote	carpals

5. Use the human biology terms to write weird sentences. For example, the word *patella* might look like this:

   "*Pa, tell a* story to us."

Here's another example:

   "Sorry, Steve. We muscle the house."
   ("Sorry, Steve. We *must sell* the house.")

ligament	saliva
flexor	nutrients
bile	alimentary

Section 24

HEALTH AND ENVIRONMENT

24–1 What Does the Sketch Represent?

The creative spirit springs into action when students attempt to match a title with a sketch.

Give students an illustration, one at a time, to examine. Then have them write what they think the illustration represents on the line below the sketch.

1.

2.

3.

4.

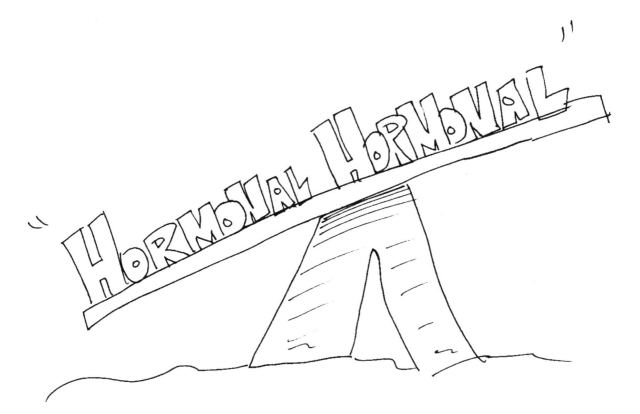

5.

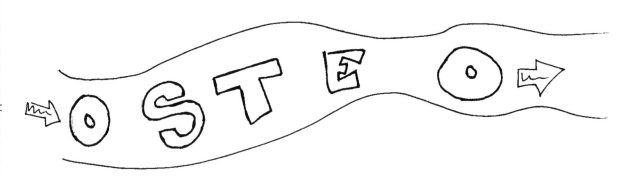

6.

7.

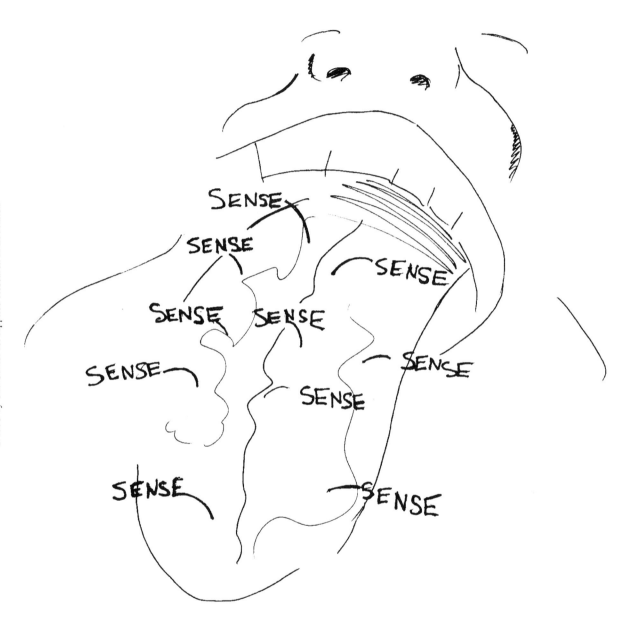

8.

Life Science

HEALTH AND ENVIRONMENT (continued)

24–2 Can You Solve the Problem?

The following miniproblems will provide a "5-minute" challenge for most of the students. Some students may be more productive if allowed to work with a partner.

1. A person *must* have certain things to stay healthy. For example, a healthy individual must have sufficient *rest*.

Use the letters in HEALTHY PERSON as the first letters to indicate what a healthy person must have. The first one is done for you.

H - hemoglobin, hormones
E -
A -
L -
T -
H -
Y -

P -
E -
R -
S -
O -
N -

2. Unscramble the six terms related to the circulatory system. Then alphabetize them and place answers in the appropriate spaces.

Scrambled Terms	Unscrambled Terms	Alphabetized Terms
• lsceaipiral	a.	a.
• saeritre	b.	b.
• nsepel	c.	c.
• ynoexg	d.	d.
• vsine	e.	e.
• tehra	f.	f.

3. The body needs organic nutrients. Carbohydrates are examples of organic nutrients. For example, simple sugars are carbohydrates.

   Make up a *five*-word description for carbohydrates. Use all 21 letters in the box to form the words.

HEALTH AND ENVIRONMENT (continued)

o	g	e	y	n	o	s
r	h	e	i	u	n	e
t	c	a	e	m	f	r

4. List *five* inorganic nutrients that have three or more vowels in their names.

5. This symbol represents something:

   Hint: Think math. The "something" has a two-word description. First, identify the two-word description. Then use the letters in the description to answer the question: What do you call a substance that causes an allergy? You may use the same letter more than once.

6. See how many times *water* appears by using the letters in the box.

h	w	h	o	e	r
w	a	r	e	t	h
o	h	a	t	h	t
w	o	h	a	e	r

24–3 What's in a Name?

Creating names from vocabulary terms provides a humorous challenge. Students have fun firing up their imaginations and producing unique responses.

1. Make up a first and last name from the eight diseases or disease-related terms.

   Example: antigen - Andy Gen

rickets	rabies
virus	malaria
jaundice	leukocytes
typhoid	microbes

2. Use the letters in each term to create a name of a *vegetable*. You may use a letter more than once.

HEALTH AND ENVIRONMENT (continued)

Term *Vegetable*

a. diabetes a.
b. carbohydrate b.
c. metabolism c.
d. nutrition d.
e. consumer e.

3. Use the letters in each term to create a name of a (an) animal(s). You may use a letter more than once.

Term *Animal(s)*

a. dermatitis a.
b. antigen b.
c. microbe c.
d. electrocardiogram d.
e. malaria e.

4. Name the five senses. The letters necessary to identify each sense are scattered in the box. You may use the same letter more than once.

g	a	f	e	r
o	i	c	h	t
n	u	m	l	s

5. Use the letters in NUTRITION to complete the name of an item, object, or event associated with health. Write each name in a vertical pattern across the word "nutrition." The first one is done for you.

```
            m
            i
    N U T R I T I O N
            e
            r
            a
            l
```

HEALTH AND ENVIRONMENT (continued)

24–4 Create-A-Comment

These two activities allow students to use humor as the motivator for creating comments.

Provide students with a copy of a drawing. Ask them to examine the drawing and write a comment or statement in the balloon next to the figure in the illustration.

1.

2.

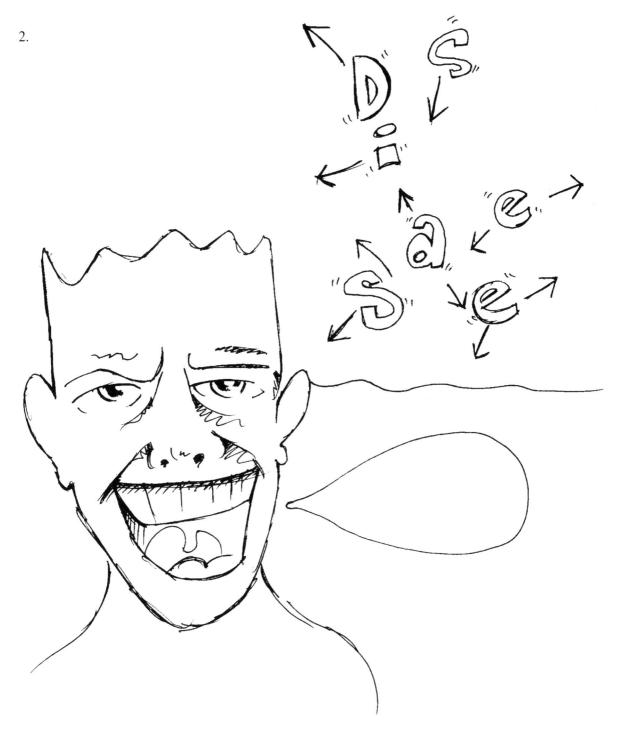

HEALTH AND ENVIRONMENT (continued)

24–5 Riddle Bits

The following 12 riddles will charge the neural batteries of students.

1. Some substances are slippery; others have rough textures. What body substance is the stickiest?
2. What are the eating habits of geometry teachers?
3. What are neuroses?
4. What is a chiropractor?
5. What respiratory illness is common among cowboys and cowgirls of the rodeo?
6. What disease is common among egotistical people?
7. What condition is caused by an overpopulation of fleas?
8. What disease is prominent among people in the roofing business?
9. What would a person receive for winning a "lack of muscle use" contest?
10. What would be another appropriate title for the fictitious Count Dracula?
11. What might a person develop if he accidentally ate a chrysalis?
12. Why did Mrs. Laria say she would never have children?

24–6 What Goes Where?

Give students the following combinations of symbols, words, and numbers to put together. Have them identify the symbols, decipher the words and phrases, and unscramble everything to reveal the correct terms.

1. Fe + O = ? (-t) + v + 👁
2. 〰️ + s.
3. ts + 🌳 + opposite of old + n.
4. 🐟 + de + opposite of out + cy.
5. o + (this way ↘) + ns + ge.
6. t + opposite of out + zero (−zo) + f + on + hero (−ho).

Life Science

HEALTH AND ENVIRONMENT (continued)

24–7 Creative Potpourri

Here are five activities for a student's creative pleasure. They are a mixture of wild words and silly definitions.

1. Match each word with a rhyming word. Try to make silly, two-word definitions or descriptions.

    Example: word: blood; matching word: flood—*blood flood* Blood flood means excessive bleeding.

    Place responses in the appropriate spaces.

	Word	*Matching Word*	*Definition/Descripiton*
a.	blood	flood	excessive bleeding
b.	sneeze		
c.	diet		
d.	calorie		
e.	fatigue		
f.	immune		

2. Create silly definitions for each of the terms. The first one is done for you.

    Example: adenoids - the accumulation of several "enoids"

Adam's apple	paranoid
hypnosis	keratin
pharynx	eardrum

3. See how many *two-letter* words you can get out of each term. You may use a letter more than once. Do not use nicknames or abbreviations.

    Example: emotion - on, in, no, to, it, and me

poison	immunity
arthritis	skeletal
adrenaline	pathology

HEALTH AND ENVIRONMENT (continued)

4. Find a science word that rhymes with each health term. Use the hint to find the correct answer.

   pathogen   *Hint:* Periodic Table of Elements
   nutrition   *Hint:* Two objects rubbing together
   immune   *Hint:* Fruit or seed of a certain plant
   disease   *Hint:* Six-legged "honey makers"

5. There are letters to *four* different health problems scattered in the box. All health problem terms begin with the *letter b*. What are they? You must use all of the letters.

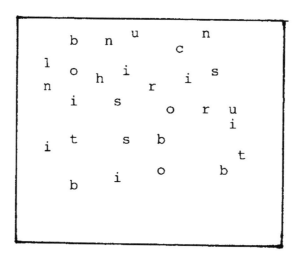

TEACHER'S GUIDE AND ANSWER KEY

PHYSICAL SCIENCE

Section 1

MEASUREMENT

1–1 What Does the Sketch Represent?

Here are some possible responses:

1. Playing "follow the liter" (leader).
2. The distance from one point to another.
3. The amount of matter in an object.
4. A parking meter.
5. Off balance.
6. Cubits - "cube its."
7. A measure of POWER.
8. The freezing point of water.

1–2 Can You Solve the Problem?

These six problems will be a measure (pun intended) of student stamina. It will also test their ability to be flexible thinkers.

1. Time - cen<u>time</u>ter.
2. M<u>ass</u>acre, m<u>ass</u>age, am<u>ass</u>, and biom<u>ass</u>.
3. 152. There are 38 animals (23 + 15). If each animal has 4 feet in a pound (animal pound, that is.)
4. Unknown. It depends on the length of the inchworm.
5. 999 ~~milli~~ liters = one liter.
6. Accuracy counts because you want to be *sure* (mea<u>sure</u>).

1–3 What's in a Name?

1. *Celsius* and the Tenth *degree; kilo, gram,* and the *Mass;* little *Kilo* and the *Meter,* and so on.
2. Telegram, grammar, gramophone, grammatical, gram positive, gram atom, and so on.

MEASUREMENT (continued)

3. Emu, mole, and mule.
4. The remaining three letters are UFN. Unscramble them and you'll see that the activity should have been FUN.
5. Ram, tiger, gnat, gar, mite, mice, rat, cat, nit, and so forth.

1–4 Create-A-Comment

Here are two possible responses:

1. "I prefer the yardstick. The meter stick covers more area."
2. "Daryl sure looks like he has put on weight."

1–5 Riddle Bits

1. Because many of them use a "liter" box.
2. It varies. If a gas and water meter are present, then you'd find two.
3. Gr<u>am</u> - *ram*.
4. M<u>ass</u> - *ass*.
5. M<u>ill</u>i - *ill*.
6. Inch by inch.
7. Because most of a <u>lite</u>r is *lite*.
8. Qu<u>art</u> - *art*.
9. Kil<u>og</u>ram - *log*.
10. Inch - *chin*.

1–6 What Goes Where?

1. gr + ees + e + d = degrees.
2. lite + ill + m + r + i = milliliter.
3. ameter + di = diameter.
4. er + me + ki + lo + t = kilometer.
5. g + i + cent + ram = centigram.
6. ale + c + s = scale.

Teacher's Key 355

MEASUREMENT (continued)

1–7 Creative Potpourri

1. 25,600 cm□ or 25,600 [cm]
2.
   scale
3. Less than one. The string of letters measures approximately 13.5 cm or 13.5 percent of a meter.
4.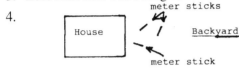
5. Te, Ti, Se, Sr, I, Re, Ir C, Cr, Tc, Cs, Ce, Sm, Er, Tm, Cm and Es.
6. MA - Master of Arts and MS - Master of Science.

Section 2

MATTER

2–1 What Does the Sketch Represent?

Here are some possible responses:

1. Gaseous matter.
2. Solid matter.
3. Matter scatter.
4. Solid shape.
5. Matter going through a *phase*.
6. Elastic matter.
7. Matter under pressure.
8. Examples of organic matter.

2–2 Can You Solve the Problem?

1. den + sity = density; hard + ness = hardness; col + or = color.

MATTER (continued)

2. decom + pose = decompose; burn +ing = burning; ru + sting = rusting.
3. tarsus - tar; monkey - monk, key; feldspar - spa; squash - ash; whelk - elk; pupa - pup, pa; saloon - loon; stalagmite - mite; shrimp - imp; travertine - tin.
4. spothtgil - spotlight; luvture - vulture; rotdne - rodent; saugase - sausage; lisver - silver; nelacdar - calendar; gargeab - garbage; emsquite - mesquite; merycru - mercury; orfceps - forceps.
5. nail - snail; rib - crib; cell - cello; hair - chair; arm - armadillo, arm chair; organ - organ grinder, organ system, organism; hip - ship, whip, hippopotamus; gum - gum drop, gumbo; liver - liverwort, liverwurst; iris - Irish setter, Irish potato.

2–3 What's in a Name?

1. Take *oil* from *soil*. Combine the remaining *s* with *wa* from *water* to form *saw*.
2. M - mothballs, mouse; A - angora sweater, ants, attache case; T - tank top, towel; T - T-shirt, tie; E - elastic headband, elastic garter; R - raincoat, rayon clothing.
3. S - seats, seatbelt; T - tires, trunk; U - universal joint, undercarriage, upholstery; F - fan, fender; F - filter, fuse.
4. a. arm + ad + ill = armadillo; b. all + at + or = alligator; c. hip + pop + am = hippopotamus; d. tar + ant = tarantula; e. whip + poor = whippoorwill; f. halo + pod = cephalopod; g. get + at + ion = vegetation; h. raw + ad = crawdad.
5. ar + ch + eo + pt + ryx = archeoptyx. The missing letter is *a*. The *a* goes between the *h* and *e;* thus the mystery animal is *archaeopteryx*.
6. A - albite, agate, aragonite; N - nickel, nephrite; Y (omit); T - talc, tourmaline; H - hematite, halite; I - ice, iron; N - nemalite, nephelite; G - gypsum, galena.
   Note: If you wish to include the letter *Y* as a "superchallenge," do so. An answer for *Y* is yellow ocher.

2–4 Create-A-Comment

Here are two possible responses:

1. "An excellent example of separating matter."
2. "Hmmm. It must be mixed matter."

2–5 Riddle Bits

1. "No problem. She's just going through a *phase*."
2. M A T T E R - "ATE" - 8.
3. Che<u>mica</u>l change - mica.
4. Six. C + H (or He) + I + Ca = 4; N + Ge = 2.

Teacher's Key 357

MATTER (continued)

5. "Fizz . . . ical."
6. A measure of matter.
7. To highlight the "proper T's" (properties) of matter.
8. Remove the i<u>qu</u>, move the <u>l</u> next to the <u>i</u>, and add <u>so</u> to the left of the <u>l</u>.
9. Four <u>thousa</u>nd pounds. 2 (to) + n = ton and 2 (to) + n = ton. Ton (2,000 pounds) + ton (2,000 pounds) = 4,000 pounds.
10. Another way of looking at matter.

2–6 What Goes Where?

1. v + ol + "you" = mmm = volume.
2. den + Si + t + y = density.
3. *First word:* period + ic = periodic; *second word:* table—*periodic table.*
4. co + m + pound + s = compounds.
5. el + e + men + t = element.
6. mole + Cu + les = molecules.

2–7 Creative Potpourri

1. T <u>U</u> R <u>K</u> E <u>Y</u> – U = uranium, K = potassium, Y = ytterium.
2. ~~MATTER~~.
3. ~~BURNING~~.
4. Three: N, S, or U. (These, of course, are chemical symbols for elements.
5. mix ——— ture or $\dfrac{\text{mix}}{\text{ture}}$ or mix/ture

Section 3

ATOMIC STRUCTURES

3–1 What Does the Sketch Represent?

Here are some possible responses:

1. The "Atoms" Family.

ATOMIC STRUCTURES (continued)

2. Atoms that make up elements.
3. A different kind of atom.
4. Electron dots.
5. The change in an element.
6. Atoms in motion.
7. Atoms on a collision course.
8. Millions of invisible atoms.

3–2 Can You Solve the Problem?

1. N - nautilus, needlefish, narwhal; U - uku (Hawaiian fish), unau (South American sloth); C - cow, coyote; L - lion, limpet; E - eagle, eel; U - ungulates (hoofed mammals), univalve (gastropod); S - sea lion, snake, snail.
2. P - pyrolusite, platinum, pyroxene; R - rhodonite, ruby, rutile; O - opal, olivine; T - tourmaline, topaz, talc; O - onyx, orthoclase; N - nickel, nephrite.
3. E - electric fan, electric iron; L - lights, lamps; E - electric clock, electricity; C - coffee maker, couch; T - table, tools; R - radio, rug; O - oven, onions; N - nightstand, nutcracker.
4. a. ant; b. bisect; c. bulb; d. tooth; e. curse; f. toupee; g. density.
5.

```
                    p
                    r
                    o
        p  r  o  t  o  n  s      Protons. Protons have a plus (+) or
                    o                       positive electrical charge.
                    n
                    s
```

3–3 What's in a Name?

1. a. Saigon; b. bonbon; c. lethargy; d. frond; e. dwell.
2. a. let; b. eel; c. less; d. select; e. sell; f. sleet.
3. Tom, Mat, Tim, Ti, and Tio.
4. Celery, corn, pea, and pepper.
5. Pineapples, dates, bananas, apples, apricots, melons, pears, passion fruit, raspberries, tomatoes, and so on.

ATOMIC STRUCTURES (continued)

3–4 Create-A-Comment

Here are two possible responses:

1. "Look, Sarah. An 'elect Ron' movement."
2. "Well, I'll be darn . . . three electron shells!"

3–5 Ribble Bits

1. From electron clouds.
2. Atomic numbers (1 = a, 20 = t, 15 = o, and so on).
3. In subshells.
4. A periodic table.
5. Separate the + into two -'s.
6. Tomato.
7. One atom - a million pencil dots.
8. They are in *reverse* orbit. Tibro spelled backward is *orbit*.
9. A male turkey - a tom.
10. NE + ON = The element NEON.
11. AMU (amu is the abbreviation for "atomic mass unit" or atomic weight).
12. A cow (cobalt = Co, tungsten = W).

3–6 What Goes Where?

1. on + t + neu + r = neutron.
2. 2,000 pounds = ton + pro = proton.
3. n + eye = i + o = ion.
4. *First word:* energy; *second word:* shell—*energy shell*.
5. 1 + 2,000 pounds = ton + DA = Dalton.
6. *First word:* nu + ear + cl = nuclear; *second word:* a few cents = change—*nuclear change*.
7. *First word:* clouds (-s) = cloud; *second word:* ham + c + ber = chamber—*cloud chamber*.

3–7 Creative Potpourri

1. *Al* would make a great *chemist;* I think we should *elect Ron;* Sharon *Ton* is a real *pro;* Shelly

ATOMIC STRUCTURES (continued)

would rather listen to the *radio* than learn about *isotopes;* Mrs. *Per* said her brother is a *cop;* Darryl said he saw a *bum* sleeping under a *plum* tree.

2. Free neutrons, no charge; protons are always positive; atoms are "out of sight"; and so forth.
3.

   matter ⇄ a t ? m ?

   The *o* and *m* are missing. Therefore, matter could not be made of atoms.
4. Stomata, tomato, stoma, toast, most, and so on. Sentences will vary.
5. Ho + Se = hose; Cu + Te = cute; Ti + Dy = tidy; La + Te = late; B + Ag = bag; Ca + Se = case; Ta + S + K = task; Pu + Pa = pupa; and so forth.

Section 4

CHEMISTRY

4–1 What Does the Sketch Represent?

Here are possible responses to the sketches:

1. A classic case of nitrogen fixation.
2. Three members of the Halogen Family.
3. The properties of fluorine.
4. Ionization (eye-on-ization).
5. The 4 C's or four "ces" or forces between mole/cules.
6. Electron (− charge) dots.
7. Electron pear (pair).
8. A crystal model.

4–2 Can You Solve the Problem?

1. CHEMISTRY
2. Turn IT into H
3. Solid, gas, liquid, oxygen, hydrogen, and transparent. The other words are charcoal, mercury, and pewter.

Teacher's Key 361

CHEMISTRY (continued)

4.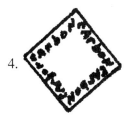

5. C - carbon, clay; H - hematite, hydrogen; E - elements, enzymes; M - matter, metal; I - ice, isotope; S - silicon, salt; T - tin, tree; R - radish, rodent; Y - yak, yogurt.

4–3 What's in a Name?

1. Methyl - Ethyl Methyl; ion - Ryan Ion; base - Mace Base; alloy - Roy Alloy; calorie - Valerie Calorie; meter - Peter Meter; mole - Cole Mole; salt - Walt Salt; weight - Kate Weight; Lye - Vi Lye.
2. Element - Ella Mint; energy - Ann Ergy; polymer - Polly Mare; molecule - Molly Cule; milliliter - Millie Leeter; generator - Gina Rader; evaporation - Eva Porasion; deliquescence - Della Quesance.
3. Gram - ram; mass - ass; catalyst - cat; concentrate - rat; constant - ant; molecule - mole; pigment - pig; and so on.
4. Sodium - sod; distillation - till; halogen - log; hydrocarbon - carbon; nitrocellulose - cell or cellulose; sedimentary - sediment.
5. Chemistry - mist; hydrogen - hydro; cloud chamber - cloud; thermal radiation - isotherm; polar bond - polar; vapor - evaporation; and so forth.

4–4 Create-A-Comment

Here are two possible responses:

1. "I just failed the Acid Test."
2. "Joe would make a good chemist. He likes to stir 'things' up."

4–5 Riddle Bits

1. A catalyst (cattle list).
2. 57 and 43 – 57 (La) + 43 (Te) = LaTe or *late*.
3. F (fluorine), U (uranium), and N (nitrogen): together they offer FUN.
4. Holmium - Ho ("Ho, Ho, Ho").
5. Sulfur (fur)
6. Antimony - the word *ant* makes up three of the eight letters or $37\frac{1}{2}$ percent.
7. In the vitriol. If you unscramble the last three letters (iol), you'll find *oil*.
8. Student B. He or she has the *right* chemistry.
9. Coleslaw (coles law).

CHEMISTRY (continued)

10. Most people might reply, "to drink."
11. Decomposition - "D" composition.
12. He felt the *oil* in b*oil*ing would lubricate the wheels.
13. The sketch shows a *hydrogen bond*.
14. U (Uranium), Ta (Tantalum), and hydrogen (H) - UTaH or Utah.
15. Mostly in three states.

4–6 What Goes Where?

1. grade + i + cent = centigrade.
2. phous + mor + a = amorphous.
3. metal + id + lo = metalloid.
4. A + TT + E + M + R = MATTER.
5. mul + for + las = formulas.
6. id + coll + o = colloid.
7. ble + ci + mis = miscible.
8. dr + ate + hy = hydrate.
9. ur + ate + at + s = saturate.
10. er + or + at + gen = generator.

4–7 Creative Potpourri

1. Fifteen: gravel + Fe; gravel + sand; gravel + S; gravel + confetti; gravel + NaCL; Fe + sand; Fe + S; Fe + confetti; Fe + NaCL; sand + S; sand + confetti; sand + NaCl; S + confetti; S + NaCl, confetti + NaCl.

2. Amalgam - a new brand of chewing gum; ketones - a singing group; iodine - the hungriest of elements (I ah dine); lignite - the opposite of "ligday"; volatile - a type of glazed ware of ornamental nature; sulfate - the future of a "sul."

3. Mixture - mixture (Mick sure) is a nice guy; metal - Roxanne *metal* (met all) of her boyfriend's band members; monatomic - *Monatomic* (Mona Tomic) was elected student body president; fungicide - Get serious. Let's put all *fungicide* (fun aside); hydride - *Hydride* (I'd ride) with Ellen to the concert on Saturday night; dynamite - "Who would ride the roller coaster? Well, *dynamite*" (Dinah might).

4. Responses will vary.

Teacher's Key 363

Section 5

FORMS OF ENERGY

5–1 What Does the Sketch Represent?

Here are some possible responses:

1. Energy changing from one form to another.
2. Light energy.
3. A ball of energy.
4. Stored energy.
5. Scattered atomic energy.
6. Heat energy.
7. Different forms of energy.
8. The measurement of energy.

5–2 Can You Solve the Problem?

1. About 67 percent of two-thirds of the energy available. Four letters in ENERGY (ENER) are needed to spell NERVE.
2. Tea, hen, and rat.
3. You can recover EERY from ELECTRICITY. Therefore, four letters out of six (ENERGY) equal two-thirds or 67 percent.
4. *First sketch:* 👁 *Second sketch:* I or aye or ay.
5. As a *hammer* drives a *nail* into the *wooden board*, the head of the nail warms.

5–3 What's in a Name?

1. Pepper, pea, and lettuce.
2. Tick and cricket.
3. Apatite, aragonite, realgar, citrine, lignite, pyrite, iron, galena, tin, and so on.
4. Magpie, crow, and parrot.
5. Syenite, granite, dolerite, shale, slate, sandstone, oolite, gneiss, and so forth.

5–4 Create-A-Comment

Here are two possible responses:

1. "Say, that's a nice 'melting point' sign."
2. "Look at that! Sound waves going through water."

FORMS OF ENERGY (continued)

5–5 Riddle Bits

1. EN<u>ERG</u>Y - ERG.
2. 12 - POT<u>IA</u>L ENERGY. If you removed *10,* you'd still have 12 letters left. PO<u>TEN</u>TIAL ENERGY.
3. About 60 percent. VIBRAT<u>IONS</u> - three letters out of five. (S<u>OU</u>ND) equals three-fifths or 60 percent.
4. MAgNETiC ENeRGY - Ag (silver) and Ti (titanium).
5. Nucl<u>ear</u> - ear.
6. H<u>eat</u> - eat.
7. <u>SOU</u>ND and <u>LIG</u>HT = SOUGHT. You'd get energy that will continue to be SOUGHT.
8. Because $\frac{4}{9}$ or 44 percent of the word po<u>ten</u>tial is a *tent.*
9. From his "Elect Ric" theme.
10. KI<u>NET</u>IC - a net.
11. Graphite - pencil lead.
12. on the m̶ove / energy (energy on the move)

5–6 What Goes Where?

1. *First word:* t + gh + li = light; *second Word:* waves—*light waves.*
2. *First word:* s + ore + d + t = stored; *second word:* er (rrr) + en (n) + "gee" (gy) = energy—*stored energy.*
3. d + *ou*nce + s = sound.
4. ice (–e) = ic + tom + a = atomic.
5. c + nu (new) + ear + l = nuclear.
6. a + ant + di + r = radiant.

5–7 Creative Potpourri

1.
```
        n
        u
        c
   s o  l  a r
        e
        a
        r
```

Teacher's Key   365

FORMS OF ENERGY (continued)

2. Pot<u>ent</u>ial energy - There is no *h*. Therefore, you can't produce *heat*.
3. MAGNETIC FIELD.
4. N + R + G + Y. Then divide the box in half according to the arrows (see sketch). Now you have  two E's. Turn the ⊰ on the right around to face east -. E Then put it all together for ENERGY.
5.

Section 6

HEAT AND ENERGY

6–1 What Does the Sketch Represent?

Here are some possible responses:

1. A bad spell of HEAT.
2. Scattered temperatures.
3. Boil*ing*.
4. Freez*ing*.
5. HEAT being absorbed.
6. Thermometer check (√).
7. Loss of HEAT.
8. A SOLID going through A PHASE.

6–2 Can You Solve the Problem?

1. TEMPERATURE = about one inch or 2.54 centimeters in length.

HEAT AND ENERGY (continued)

2. Hang a thermometer from the patio roof or front door. Then you can measure the temperature in the front and back yards.
3. And the letter C to HEAT. Now you have CHEAT. That's how.
4. Fahrenheit = ahet or heat. Forty percent of Fahrenheit produces heat.
5. <u>heat</u>   h↘e↙at   (heat)
6. Kinetic energy.

6–3 What's in a Name?

1. H - <u>h</u>umidity, <u>h</u>umid, <u>h</u>ygrometer; E - evaporat<u>e</u>, condens<u>e</u>, pressur<u>e</u>; A - he<u>a</u>t, we<u>a</u>ther, evapor<u>a</u>tion; T - a<u>t</u>mosphere, s<u>t</u>ratus, s<u>t</u>ationary front.
2. No, not, note, noon, nine, nit, neon, nice, none, and so on.
3. Cat, rat, ape, asp, ass, eel, and so forth.
4. He hit Fran.
5. <u>Temper</u>ature - temper; pressure <u>cook</u>er - cook; molecular <u>act</u>ion - act; and so forth.

6–4 Create-A-Comment

Here are two possible responses:

1. "Notice the small change in the TEMPERATURE."
2. "I guess they couldn't stand the heat."

6–5 Riddle Bits

1. Only WR. Three-fourths of HEAT would be EAT in WATER. Therefore, only WR remains.
2. EAT, ATE, or TEA. All three produce HEAT.
3. HE ATE TEA or HE ATE A HAT.
4. Light energy.
5. <u>End</u>othermic reaction. The first part "end" suggests termination.
6. Convection has a *ve* combination for the fourth and fifth letters. Conduction has a <u>du</u> combination for the fourth and fifth letters.
7. Remove the *burn* from burning.
8. It will never make a "fuel" out of itself.
9. A "ula" - ins*ula*tor.

Teacher's Key 367

HEAT AND ENERGY (continued)

10. Because of the *bus* in combustion.
11. Chemical reaction - hemlrat: *thermal*.
12. They're covered with too many scales.
13. Arm - *warm*.
14. Because of their *temper*.

6–6 What Goes Where?

1. *First word:* a + sol + ute + b = absolute; *second word:* 0 = zero—*absolute zero*.
2. in or en + fair or Fahr + height = *Fahrenheit*.
3. see + cell + us or si + cel + us = *Celsius*.
4. d + a + tion + Ra + i = *radiation*.
5. ot + ion + m = *motion*.
6. tea + m + s = *steam*.

6–7 Creative Potpourri

1. Convection.
2. Radiation.
3. Melt, hot, thermal; 100, 212.
4. Only *two*. It doesn't matter how many t's or e's come from the box. There are only two a's.
5. molecules on the move

Section 7

FORCE AND MOTION

7–1 What Does the Sketch Represent?

Here are some possible responses:

1. Balanced "four-ces" (forces).
2. Net force.

FORCE AND MOTION (continued)

3. Forces in motion.
4. The center of gravity.
5. Constant velocity.
6. Falling objects.
7. Total force.
8. Backward motion.

7–2 Can You Solve the Problem?

1. EQUIL - I - $\overrightarrow{\text{BRIUM}}$ (with leftward arrow under EQUIL)
2. $\text{FORCE} \rightarrow$
   $\leftarrow \text{FORCE}$
3. ↑ ful<u>c</u>rum ↑ (with fulcrum symbol)
4. Work = force times distance.
5. NEW + 2,000 LBS (ton) = NEWTON.
6. M(OMEN)TUM

7–3 What's in a Name?

1. Inertia, friction, acceleration, motion, projectile, reaction, kinetic, potential, and so on.
2. Velocity - Vella Siti; inertia - Ena Shur; kinetic - Ken Atic; mechanical - Mick Anickle.
3. Answers will vary.
4. Momentum - "Is this the *moment, um?*"; thrust - "This is the four*th rusty* nail I found"; Newton - "Say, that's a *new tone,* isn't it?"; heat - S*he at*e the whole pie by herself.
5. Rate - the number that follows seven; weight - the number that comes before seven; mass - six pounds of mud in a four-pound bag; speed - a word that rhymes with seed or heed or read or tweed or. . . .

7–4 Create-A-Comment

Here are two possible responses:

1. "See, Julia. Forces *do* come in pears (pairs)."
2. "Say, John really does accelerate."

FORCE AND MOTION (continued)

7-5 Riddle Bits

1. Partial vac<u>uu</u>m.
2. d - accelerate<u>d</u>.
3. Eight - w<u>eight</u>.
4. In uniform motion.
5. Every force has its <u>moment</u>.
6. The first five letters in the word.
7. Because about 43 percent of its contents are *rum*.
8. When they say with a stern voice, "Come here this moment."
9. Increased friction.
10. Both terms have two e's.
11. Backward <u>force</u>.
12. Accele<u>rate</u>.

7-6 What Goes Where?

1. en + tum + mom = momentum.
2. lc + lc + lc + lc = 4 c's or *forces*.
3. lo + city + ve = velocity.
4. w + eight = weight.
5. *First word:* negative; *second word:* ration + ccele + a = acceleration—*negative acceleration*.
6. *First word:* circular; *second word:* motion—*circular motion*.

7-7 Creative Potpourri

1. 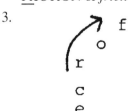 This placement produces a RUB; hence RUB produces friction.
2. Francium + iodine + carbon + titanium + oxygen + nitrogen = Fr + I + C + Ti + O + N = F<u>r</u>ICTiON or *friction*.
3.
```
        ↗ f
       ⌒  o
       r
         c
         e
```

FORCE AND MOTION (continued)

4. Write the words *air resistance* with a pencil. Now take an eraser and erase the words. There. You've done it!
5.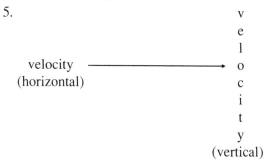

Section 8

MACHINES

8–1 What Does the Sketch Represent?

Here are some possible responses:

1. An inclined plane.
2. The "main" (mane) source of horsepower.
3. Effort arm versus resistance arm.
4. Efficiency (a fish in sea).
5. Block and tackle.
6. "Pull . . . EY" or pulley.
7. Reduced friction.
8. Thin wedge.

8–2 Can You Solve the Problem?

1. M<u>A</u>CHINE<u>S</u> - EAS, no Y; therefore, EASY doesn't exist.
2. s - screw; i - inclined plane; p - pulley; l - lever.
3. W P <u>O</u> U <u>R</u> K T. An example of WORK in PUT.

MACHINES (continued)

4. In (indium) + Cl (chlorine) + I (iodine) + Ne (neon) = the "compound" InClINe or incline.

5. M<u>A</u>CHINE MACHINE CHINE

8–3 What's in a Name?

1. M<u>A</u>CH<u>I</u>N<u>E</u> - ACHE.
2. EVE and LEE.
3. Force - horse; wheel - teal; bar - gar; torque - stork; plane - crane; incline - swine.
4. Wedge, pulley, and lever.
5. FOOT - walking; IRON - pressing clothes; TON - 2,000 pounds of weight; COT - the legs resting against the floor; RIFT - a shallow place in a stream.

8–4 Create-A-Comment

Here are two possible responses:

1. "Mr. Teeter makes a super fulcrum, doesn't he?"
2. "Melvin is using a third-class lever to destroy a first-class lever."

8–5 Riddle Bits

1. The letter <u>w</u>.
2. A fixed pulley.
3. Lever<u>age</u> - age.
4. Gene MacHine
5. Because it is a <u>first-class</u> lever.
6. Either le + ver or lev + er.
7. MA.
8. C<u>ompound</u> machine.
9. The <u>c</u>: <u>e</u> f f <u>o</u> r t.
10. A simple machine. A combination wheel and axle.
11. A <u>chin</u> is a front part of a lower jaw and chin makes up half the letters in ma<u>chin</u>es.
12. Only one. Sla<u>nt</u>ed ramp.

MACHINES (continued)

8–6 What Goes Where?

1. edge + w = wedge.
2. chin + ma + e = machine.
3. eve + r + l = lever.
4. fish ("ffic") + e + ien + cy = efficiency.
5. *First word:* heel + w = wheel; *second word:* l + e + ax = axle—*wheel* and *axle*.
6. fric + tion = friction.

8–7 Creative Potpourri

1. (across) bicycle; (down) scissors.
2. = REF (short for Referee).
3.
4.

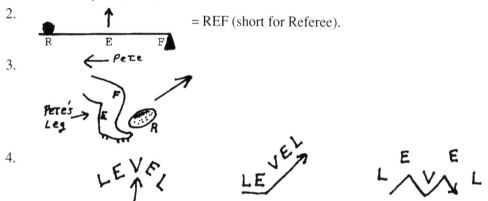

5. force course, splice device, spoon boom, fork torque, bully pulley, and great weight.

EARTH SCIENCE

Section 9

ENERGY SOURCES

9-1 What Does the Sketch Represent?

Student responses will vary. Here are some possible answers for each sketch:

1. Low on energy.
2. Energy building.
3. Solar heat.
4. Black gold.
5. The four stages of coal development.
6. Solar cell.

9-2 Can You Solve the Problem?

1. A nervous tic (Energ<u>tic</u>)
2.
```
        b                   l
        i                   i
        t                   g
        u                   n
        m                   i
        i                   t
      a n t h r a c i t e
        o
        u
        s           p e a t
```

   Peat is *not* connected with the other stages of coal development.

3. WINDMILL: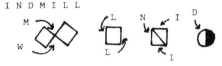

4. Shade in the three terms PEAT, COAL, and OIL. Once the peat, coal, and oil are obliterated, there will be "O" or *nothing* left.

5. There are enough letters to spell OIL 11 times *if* you count the *large* "O" surrounding the other letters.

373

ENERGY SOURCES (continued)

9–3 What's In a Name?

1. Rat, eel, goat, hog, teal, llama, and so forth.
2. Puffin, duck, skunk, and so on.
3. Toil - oil; linoleum - petroleum; dynamite - lignite or anthracite; luminous - bituminous; friend - wind; mule - fuel; fleet - peat or heat; mole - coal; polar - solar; seam - steam; bass - gas; illusion - fusion; decision - fission; sunflower - hydropower; daughter - water.
4. F - ferns; O - overlying pressure, organic; S - swamp, seed ferns, sediments; S - sands, shales, sedimentary rock; I - inland basin, interior basin or inland sea; L - lignite.
5. Methane (CH_4) and Butane (C_4H_{10}).
   a. No *two* hydrocarbon formulas on the list can be made from an H (15) and C (5) combination.
   b. No *two* hydrocarbon formulas on the list can be made from an H (14) and C (6) combination.
6. Petroleum, natural gas, and coal.

9–4 Create-A-Comment

Keep in mind that student response may be completely different from these answers. This should pose no problem since the illustrations suggest open-ended responses. Remind students to seek comments or statements with a link to the subject of energy.

1. "They seem to be running out of ENERGY."
2. "Because it's his pet. Get it? Pet - petroleum. Pretty dumb, huh?"

9–5 Riddle Bits

1. Both words contain "amps."
2. L<u>ignite</u>.
3. They were under pressure to do so.
4. The letters ROLE play a ROLE.
5. In oil pools.
6. You could add <u>con</u> to <u>fusion</u>, thus producing *confusion*.
7. Six thousand people liked it, six thousand people didn't like it. There was an even number of pro's and con's: PROCON. *Note:* This, of course, is a hypothetical study.
8. S<u>team</u>.
9. Because three out of the four letters in WIND (WIN) suggest positive results.

Teacher's Key

ENERGY SOURCES (continued)

10. Because THE SKY does not have the letters o, i, and l in their spelling.
11. Crude oil.
12. "Hi" energy (high energy).

9–6 What Goes Where?

1. out (in) + d + w = wind.
2. eat (ate) + r + w = water.
3. d + s + sick (ill) + win + m = windmills.
4. (second word) ring + s + sp = springs (first word) hot: *hot springs*.
5. (second word) cell (first word) r + sol + a = solar: *solar cell*.
6. (first word) s + bio + ma + s = biomass (second word) l + fue + s = fuels: *biomass fuels*.
7. (first word) w + p + er + o = power (second word) plant: *power plant*.

9–7 Creative Potpourri

1. Answers will vary.
2. oil foil - a fencing contest; fission fishen' - a fishing contest for nuclear scientists; fuel duel - two people fighting with lighted torches; solar bowler - an outdoor summer bowling tournament; wind spin - working a tightrope in a wind tunnel; peat feat - running a foot race in a peat bog.
3. Tar - *g*ar and shale - *w*hale.
4. Six COAL'S and the letters spelling *six times* should be darkened. The unshaded letters spell COAL.
5. One possible solution is as follows:

```
            l           a
       b i t u m i n o u s
            g           t
            n           h
            i           r
    p e a t             a
            e           c
                        i
                        t
                        e
```

6. Answers will vary.

Section 10

ROCKS AND MINERALS

10–1 What Does the Sketch Represent?

Student responses will vary. Here are some possible answers for each sketch. The hints for sketches 2, 3, 5, and 6 will help direct student thinking.

1. Crystal ball.
2. Malachite.
3. A mountain out of a molehill.
4. From Mud to Shale to Slate: the best double play combination in the history of metamorphic baseball.
5. "Silly-Cats" (silicates).
6. "Floor-ite" (fluorite).
7. Lead crystal.
8. The mineral graphite (*graph . . . ite*).

10–2 Can You Solve the Problem?

1. LEAD → Au. Make the symbol for GOLD, Au, out of the letters A and D from LEAD. Turn the letter D on its "back" (D → ⌣) and remove the line (⌣ → ⌣).
2. The child was considered a member of the CLAN: chlorine = CL and sodium = Na. However, Mr. Sodium was "backward." Therefore, Na becomes aN. So together we have CL + aN or CLAN.
3. Use the chemical symbols of the three elements to complete the word ROCK: oxygen - O, carbon - C, and potassium - K.
4. $\frac{\text{tremendous pressure}}{\text{rock}}$
5. Shale → slate. The *h* in *shale* flattens into the *l* in *Slate*. Then *l* in *shale* is turned upside down and slightly flattened on top to form the *t* in *slate*.
6. Letters c, e, f, and h have the missing i's: DIORITE, PERIDOTITE, MICROCLINE, AND LIMONITE. The remaining words are the following: a. FELSITE; b. DIABASE; d. GYPSUM; g. PYROXENE.

10–3 What's in a Name?

1. Feldspar - par, pa, spa; hematite - ma, mat; serpentine - pen, pent, tin; sedimentary - dime, dim, tar; anticline - ant, tic, line; sandstone - stone, ton, tone, sand; conglomerate - con, rate, rat; quartzite - quart, art, zit; dolomite - omit, mite; and syenite - yen, nit.

ROCKS AND MINERALS (continued)

2. *Minerals:* tz - quartz; lc - talc or calcite; yp - gypsum; ph - amphibole; nn - cinnabar. *Rocks:* tz - quartzite; bb - gabbro; yr - porphyry; cc - breccia; ss - gneiss.
3. Wolframite (wolf or mite) - the animal's name is in the word; chromite (mite) - the animal's name is in the word; sulfur (fur) - an animal's coat or covering; realgar (gar) - the animal's name, a fish, is in the word; serpentine (serpent) - the animal's name is in the word; stilbite (bite) - an animal that bites.
4. Brown coal - row; cassiterite - sit; clay - lay; feldspar - spar; corundum - run; pegmatite - peg; plagioclase - lag; travertine - rave; uranium - ran.
5. a. agate, C; b. realgar, D; c. selenite, F; d. pumice, E; e. basalt, B; f. gneiss, A.
6. R - rutile, realgar, rhodonite, rhodochrosite; O - opal, orthoclase, oligoclase; C - chromite, chalcopyrite, cryolite, calcite, cuprite; K - kaolinite, kernite, kyanite.
7. M - marcasite, malachite, magnesite, microcline; I - indicolite, iodyrite; N - nephelite, natrolite, nemalite; E - epidote, enstatite, essonite, erythrite; R - rubellite, rhodocrosite, rhodonite, roscoelite; A - albite, anorthite, andesine, apatite, amphibole; L - labrodorite, leucite, lepidolite, limonite, lazurite.
8. Felsite, diorite, syenite, and granite,
9. Nail *shale* pail; rate *slate* gate; list *schist* mist; peace *gneiss* lease; garble *marble* warble; rough *tuff* stuff; talk *chalk* stock; flirt *chert* skirt.

10–4 Create-A-Comment

Here are two possible responses:

1. "Joe talks like he's got rocks in his head."
2. "Rats. Look at that! Another rocky road ahead."

10–5 Riddle Bits

1. Fool's gold.
2. Probably between a rock and a hard place.
3. Gypsum (gyp . . . some).
4. Fluorite (fluo . . . rite).
5. Apatite (appetite).
6. Quartz (qu*artz*).
7. Feldspar (feld*spar*).
8. To a monoclinic (Mr. Crystal described a monoclinic crystal.).

ROCKS AND MINERALS (continued)

9. Tourmaline (tour . . . maline).
10. Olivine (olive . . . ine).
11. Metamorphic (met-a-morphic).
12. Concretions (concret . . . ions).

10-6 What Goes Where?

1. Pyroxene.
2. Igneous.
3. Pyrite.
4. Crystallize.
5. Gneiss.
6. Stone.
7. Serpentine.

10-7 Creative Potpourri

Answers will vary. Here are some possible responses:

1. "No, thanks. I'll take it for granite (granted)."
   "Is Carla your gneiss (niece)?"
   "Do I want to go? You breccia (betcha)."
   "Don't worry, Len. I'll let you in prophyry (for free)."
   "Bill would give you the chert (shirt) off of his back."
   "Chal (Cal) ced (said) ony (any) body can go."
2. Answers will vary, but these are possible answers: marble - cigarette brand; stalagmites - cave insects; agate - movable barrier; sediments - candy "sedi's."
3. Answers will vary, but here are some possibilities: "Caught between a rock and a hard place," "Hit rock bottom," "Head full of rocks," and so on.
4. Responses will vary.
5. Responses will vary.

Teacher's Key 379

Section 11

VOLCANOES AND EARTHQUAKES

11-1 What Does the Sketch Represent?

Here are possible answers for each sketch:

1. Magma rising or rising magma.
2. Ring of fire.
3. P wave.
4. S wave.
5. Landslide.
6. Turbulent or unsettled crust.
7. Island chain.

11-2 Can You Solve the Problem?

1. Lagma - lava and magma; truption - tremor and eruption; epicus - epicenter and focus; badrock - basalt and bedrock.
2. Jumpy, shaky, jittery, "crazy," and so on.
3. *Si*dney (silicon - Si) and *Ca*rla (*ca*lcium - Ca).
4. You can find enough letters if you use the ⌒ figure for the missing *M*. The figure represents a volcano. *Note:* The missing *M* isn't exactly inside the figure, but since *it is* the figure, both inside and out, we can win on a technicality.
5. Obsidian, pumice, and andesite.
6. Volcanoes.

11-3 What's in a Name?

1. As*theno*sphere - hen and Enos.
2. Krakatoa, Mt. St. Helens, Kilimanjaro, Paricutin, Stromboli, Kilauea, Surtsey, and Vesuvius.
3. Rock fracture, rock shock, rock snap, rock pop, rock rip, and so on.
4. a. Bogus focus; b. basalt assault; c. vibration gyration; d. epicenter renter.
5. Moho, core, tremor, and so forth.

VOLCANOES AND EARTHQUAKES (continued)

11–4 Create-A-Comment

Here are two possible responses:

1. "Jerry's *seismo* - graph joke is getting old."
2. "Oh. Oh. Another eruption warning sign."

11–5 Riddle Bits

1. Alfred Wegener (1915).
2. The focus of attention.
3. Conditions change from rapture to rupture.
4. Tectonic plates.
5. Ant (mantle).
6. Crater Haters.
7. Not to find any *fault* in his or her work.
8. Tremor over the rocks.
9. Shaker maker.
10. Spew pu.
11. Lava cone.
12. They both experience foreshock.
13. Between the p and the n - epicenter.
14. Bueno Volcano or Volcano Bueno.
15. Spent vent.

11–6 What Goes Where?

1. c + tin + ex + t = extinct.
2. m + ic + eyes (eis) + s = seismic.
3. am + sun + t + i = tsunami.
4. s + canoe + ol + v = volcanoes.
5. r + r + eight (ate) + c = crater.
6. t + fall (faul) = fault.
7. c + k + four (fore) + sh + o = foreshock.

VOLCANOES AND EARTHQUAKES (continued)

8. waves + ear + sh = shear waves.
9. ry + eye (i) + ma + pr + waves = primary waves.
10. s + m + tree (tre) + or = tremors.

11-7 Creative Potpourri

1. Responses will vary.
2. Vent his/her anger," "Let off steam," "Ready to explode," "Blowing off a lot of smoke," "Erupt with anger," and so on.
3. SCARP - scar, car and carp. The carp with the scar drove the car.
4. Responses will vary. Possible answers:

A naive person thinks . . .

bedrock is a hard place to sleep.

an *epicenter* is a place where "epi's" hang out.

Krakatoa is the result of landing hard on your feet.

crater is the opposite of lesser.

scoria is when a game player gets the winning "ia."

a *tremor* does landscaping work for a living.

Section 12

FOSSILS AND GEOLOGIC TIME SCALE

12-1 What Does the Sketch Represent?

Here are some possibilities:

1. Radioactive decay.
2. Dated rocks.
3. Division of time.
4. The Mesozoic Era follows the Paleozoic Era.
5. Carbon-14.

FOSSILS AND GEOLOGIC TIME SCALE (continued)

6. Trilobites.
7. Carbon dating.
8. Ice Age.

12-2 Can You Solve the Problem?

1. Toe, nose, senses, dentine, intestines, and so on.
2. Strata, fossilization, preservation, petrification, carbon-14 dating, sedimentation, carbonization, correlation, quaternary, coelenterates, invertebrates, vertebrates, relative age, and so on.
3.

    Teacher's Desk

13.8	13.9	14	14
14.1	14.6	14.7	14.9
14.9	14.9	15	15.1
15.2	15.2	15.3	15.3
15.4	15.5	15.6	15.8

4. Cenozoic: humans appear, mammals abundant; Mesozoic: first birds, dinosaurs dominant and dinosaurs extinct; Paleozoic: trilobites abundant, first fish, first land plants, first reptiles, and first amphibians.
5. O → OR → ORG → ORGA → ORGAN → ORGANI → ORGANIC
6.
    ```
            s
    g e o l o g i c
            c
            t
            i
            o
            n
    ```
7. "Equilibria."
8. A - ant, aphid; M - moth, mosquito, mite; B - bee, beetle, butterfly; E - earwig, embiids; R - roach, robber fly.
9. 23   A FOSSIL

12-3 What's in a Name?

1. Dee and Kay (decay).
2. Dinah (dinosaur).

Teacher's Key

FOSSILS AND GEOLOGIC TIME SCALE (continued)

3. Amber.
4. Abe (s<u>abe</u>r-toothed cat).
5. Leo (p<u>aleo</u>cene).
6. Sig Fig; Sam or Pam Clam; Lee or Dee Bee; Dale or Gale Snail; Nick, Dick, Mick or Rick Tick; Tish Fish; Di, Vi or Si Fly; Clem or Em Stem.
7. Paleozoic (<u>Paleo</u>zoic).
8. Penny, Annie, Sylvia, Anna, Anne, Ann, Lea, Nina, Ella, Lane, Lana, Sally, Ally, Rae, Erin, Rena, Diana, Diane, and so on.

12–4 Create-A-Comment

Here are two possibilities:

1. "My favorite exhibit: The Age of Dinosaurs."
2. "Wow! Look! A Headlessaurus."

12–5 Riddle Bits

1. An impression doing an impression.
2. An example of a Guide Fossil.
3. A scallop (pecten). The ribbed parts of the shell are called ears.
4. The seven periods of the Paleozoic Era.
5. Mammoth (<u>mammoth</u>).
6. Molds (m<u>olds</u> . . . <u>olds</u>mobile).
7. Rat (verteb<u>rat</u>e).
8. Dion and Leon Eon.
9. Kara and Sarah Era.
10. Because paleontologists date rocks.
11. By having a group of letters that spell EON.
12. In a geologic column.
13. Petrified (<u>petri</u>fied).
14. Eight (FOSSIL B___ ___ ___ S = 8).

FOSSILS AND GEOLOGIC TIME SCALE (continued)

12–6 What Goes Where?

1. nt + dime + se = sediment.
2. on + e = eon.
3. period + ssss + iiii + pp + man = (Mississippian period).
4. tiles + rep = reptiles.
5. s + ca + t = cast.
6. Half-life.
7. w + amp + s = swamp.
8. Na + quater + ry = quaternary.

12–7 Creative Potpourri

1. Responses to pun statements will vary.
2. *Possible answers:* Sandy strata and the Imprints; Lilly and the Crinoids.
3. *Possible answers:* crinoid - an unhappy "noid"; strata - a stratagem that came up short; guide fossils - Mr. Fossils, a sight-seeing tour guide; trilobite - the teeth marks of an angry "trilo"; nautilus - a famous knot named after James Ilus; extinction - when Mrs. Tinction divorces Mr. Tinction.
4. Answers will vary.
5. Slogans will vary. *Possible responses:* Paleontologists SOIL their clothes. Fossils are old news.
6. Words will vary. *Possible responses:* cold mold, dry fly, mud bug, and so on.

Section 13

THE EARTH'S FORCES

13–1 What Does the Sketch Represent?

Student responses to these activities will vary. There may, in some cases, be several excellent responses that match a particular sketch. Here are some possible comments:

1. Pressure buildup.

THE EARTH'S FORCES (continued)

2. Reverse fault.
3. High temperature and pressure.
4. Cooling rate.
5. A force acting on a solid.
6. Tensions (10 "sions").
7. A mountain undergoing an uplifting experience.

13–2 Can You Solve the Problem?

1. Tension, compression, and shearing.
2. Heat.
3. a. shearing; b. compression; c. tension.
4. F - folding, faulting; O - overthrust, overturning, orogeny; R - reverse faulting, Rocky Mountains, recumbent fold; C - cretaceous, Circum - Pacific Belt; E - erosion, eruption, exfoliation.
5. Various forces (four "ces").
6. Folded and faulting.
7. $\frac{\text{stress}}{\text{Earth}}$.
8. Convection current.

13–3 What's in a Name?

1. Heat feet, strain brain, metamorphic (endo-, ecto-, or mesomorphic), compression digestion, twist wrist, and tear hair.
2. Hip and arm.
3. P̲r̲essure, h̲e̲a̲t̲, and ten̲s̲ion = stress.
4. Grand Tetons and Sierra Nevadas.
5. a. faulting; b. dome mountains; c. folding.
6. Weathering.

13–4 Create-A-Comment

The nice thing about creating a comment or statement is the open-ended responses students make. A student should receive credit for any response related to the Earth's forces theme. Here are two possibilities:

THE EARTH'S FORCES (continued)

1. "Yes, a definite example of folding under pressure."
2. "Sure. A de . . . formation."

13–5 Riddle Bits

1. Both items belong in a stress test.
2. Four are left. Rock numbers 2, 3, 4, and 5 are *left* of the hill.
3. Nobody's fault.
4. Easy. Cross out the double letters *r, s,* and *e* in the word PRESSURE, and the remaining letters are *p* and *u,* which stands for *pu.*
5. Both have gone through the folding process.
6. A fault-block.
7. It takes a certain force to tilt them.

13–6 What Goes Where?

1. in + g + old + f = folding.
2. ma + g + ma = magma.
3. ol + ten + m = molten.
4. rain + t + s = strain.
5. tilt + in + g = tilting.
6. sss + tre = stress.
7. in + ear + g + sh = shearing.
8. si + ten + on = tension
9. f + li + up + t = uplift.
10. cl + i + ine + ant = anticline.

13–7 Creative Potpourri

1. Folding.
2. Answers will vary. *Possible responses:*
   David got tired standing at *tension.*
   Barbara said, "I need to board this *strain.*"
   Bob told his mother that he'd *heat* something later.
   Jean said her arm *fault* sore.

Teacher's Key 387

THE EARTH'S FORCES (continued)

Paul's *shearing* seems to be getting worse.

Thrust was eating through the pipes.

3. *Across:* tension, heat, strain; *Down:* shearing.
4. a. heat; b. pressure; c. strain; d. tension.
5. Scientists, movement, lithospheric plates. Answer to question: convection currents.
6. crate - plate; motions - oceans; toil - soil; fountains - mountains; peaches - beaches; must - crust; shivers - rivers; thrills - hills; nation - vegetation.
7. Uplift, compression, heat, fault, and strain.

Section 14

WEATHER AND CLIMATE

14–1 What Does the Sketch Represent?

1. Out in the elements or surrounded by the elements.
2. Tornado (torn . . . ado).
3. Cyclones (cyclones).
4. A bad spell of weather.
5. Weather Watch.
6. Barometer.

14–2 Can You Solve the Problem?

1. a. Ar + Ir − r = air.
   b. N + I + Ra = rain.
   c. O + Sn + W = snow.
   d. S + Ga = gas.
2. a. rain.
   b. snow.
   c. hail.
   d. sleet.

WEATHER AND CLIMATE (continued)

3. *Possible answers:* recipe, can, pepper, pot, potato, carrot, tea, pan, ice, onion, rice, pie, receipt, cat, and tape.
4. There is no way. The words *the atmosphere* do not have an i or u—m<u>oi</u>s<u>tu</u>re.
5. *Possible answers:* C - cold, Celsius; L - latitude, land; I - ice, icicle; M - moisture, marine, maritime; A - air, absorption; T - temperature, tropical; E - evaporation, energy.
6. A recipe for atmosphere.

14–3 What's in a Name?

1. *Possible answers:* Norm Storm, Moe/Joe/Flo Snow, Wayne/Jane Rain, Dale/Gayle Hail, Sue/Lou Dew, Si/Vi High, Bo Low, and Pete Sleet.
2. *Possible answers:* anemometer - Anna Mometer; millibar - Millie Barr; aerosals - Ara Salz; radiation - Ray De Ashon; aneroid - Anne Aroyd; ionosphere - Ian Osfeer; Fahrenheit - Farron Hite; radiosonde - Ray Deoson; maritime - Mary Tyne; cyclone - Si Klone; hurricane - Harry Kane; altostratus - Aldo Stradis.
3. Answers will vary.
4. *Possible answers:*
   a. Four terms with two i's: inversion, maritime, radiation, millibar, lightning, and adiabatic.
   b. One term with three o's: monsoon.
   c. Four terms with two a's: radiation, evaporate, saturation, radar, stationary, aerovane, and rainfall.
   d. Five terms with two o's: typhoon, hydrologic, tornado, convection, conduction, hydrologic, ionosphere, and troposphere.
   e. Three terms with two r's: thermometer, barometer, cirrus, thermograph, and hygrometer.
5. *Possible answers:* meter, log, toy, room, tee, yoyo, root, letter, leg, gem, meteor, loot, and so on.

14–4 Create-A-Comment

Provide each student with a copy of a drawing, such as "Flying Cups?" Call attention to three or four words that you suggest are related to a theme—in this case weather. Tell the students that the words *mug, soup, fog,* and *pour* will act as motivational triggers to coax their creative spirit. They are to think of creative ways in which the words relate to the theme and to the drawing.

Flying cups? Possible statements: "It sure is a muggy day." "Nuts. Here comes that soupy fog again." "Get ready. It may pour any minute."

Teacher's Key

WEATHER AND CLIMATE (continued)

It makes cents key words: weather, air, and cents. *Possible statements:* "Do you notice the change in the weather?" "There's definitely change in the air." "This crazy weather makes little 'cents'." Before students make any sense out of cents, tell them it's okay to play with words.

14–5 Riddle Bits

1. Nim<u>bus</u>.
2. Because they make excellent hare (hair) hygrometers.
3. Leeward (Lee and Ward).
4. Argon (R going, going, gone!).
5. Thin air.
6. Dol<u>drums</u>
7. A break in the weather.
8. Trade Winds.
9. Horse Latitudes.
10. C<u>loud</u>.
11. Middle latitudes.
12. Forecaster.
13. Relative humidity.
14. Eyes (ice) over an area.
15. Thunder Downunder.

14–6 What Goes Where?

1. one + Oz = ozone.
2. aves + w = waves.
3. g + light + in + n = lightning.
4. ons + ons + o + m = monsoons.
5. rated + saturn − rn = saturated.
6. eye (i) + ill + bar + m = millibar.
7. sun (son) + sea = season.

WEATHER AND CLIMATE (continued)

14–7 Creative Potpourri

In Exercises 1–5, answers will vary.

5. *Possible answers:* a. mass/mother - Half the letters in mass refer to mother (ma).
   b. A lion will find a den in condensation.
   c. There doesn't seem to be any ease among the "eee's" in freeze and sneeze.
   d. They both are involved with rent.
   e. Both of them rely on the "other."
6. Answers will vary.

Section 15

ASTRONOMY

15–1 What Does the Sketch Represent?

Student responses will vary. Here are some possible responses for the sketches:

1. Quarter moon.
2. The Earth's "axes."
3. A light-year (one day short).
4. Big-bang theory.
5. Full moon.
6. Falling star.
7. Sunset.

15–2 Can You Solve the Problem?

1. Mars, Saturn, Uranus, and Neptune.
2. Planet, comet, moons, and meteors.
3. Chromosphere, corona, and photosphere.
4. Aurora borealis.
5. 88,325 Earth-days.

Teacher's Key

ASTRONOMY (continued)

6. It's an example of a partial eclipse.
7. Solar paneling.

15–3 What's in a Name?

1. A<u>nnular</u>; st<u>ell</u>ar or <u>stell</u>ar; co<u>ron</u>a; <u>sol</u>ar; <u>planet</u>; r<u>adar</u>.
2. Sally Halley, Mona Corona, Lars Mars, Michael Cycle, Nora Aurora, Janice Uranus, Mace Space, Dave Wave, Brian Orion, and June Moon.
3. Sky - thigh or eye; space - face; apogee - knee; heat - feet; flare - hair; eclipse - lips or hips; rocket - socket; gravity - cavity.
4. Fly - sky; loon - moon; mite - light or meteorite; gar - star, dalmation - constellation; zebra - Libra; polar bear - solar flare; musk ox - equinox.
5. Astro - gastro; light - blight; ursa - bursa; air - hair.

15–4 Create-A-Comment

1. *Possible response:* "Look! You can see the Big Zipper."
2. *Possible response:* "Wow! The constellation 'Rubbing Stars.'" A superexample of science friction.

15–5 Riddle Bits

1. Zenith.
2. Satellites.
3. Jupiter.
4. Because of the fly traps.
5. Moon (mooo . . . n).
6. Mercury.
7. Neptune.
8. Saturn.
9. Antares.
10. Ju<u>pit</u>er (a pit).
11. Numb. The middle of a pe<u>numb</u>ra is *numb*.
12. Mars. It has a polar cap.
13. WONDERS of the heavens.
14. Because it has a ring around it.

ASTRONOMY (continued)

15–6 What Goes Where?

1. E + LIPS + E L or ELLIPSE.
2. I + VERSE + U N or UNIVERSE.
3. RAT + E R + C or CRATER.
4. FRACTION + R E or REFRACTION.
5. S + TURN + A or SATURN.
6. A R + E S + ANT or ANTARES.
7. BED + O + A L or ALBEDO.
8. *First word:* MA + P + ROCKS + I or PROXIMA; *second word:* R I + CENT + Au or CENTAURI.
9. VA + NO or NOVA.
10. T + EAR + H or EARTH.

15–7 Creative Potpourri

Here are some possible responses to share with students:

1. a. The sun is a powerful force in the universe.
   b. Jupiter is the largest Jovian planet in the solar system.
   c. The Newtonian telescope is a reflecting telescope.
2. *Possible answers:* solar, star, shine, spectrum, sunset, sunrise, sunspots, streamers (prominences), sundial, solar flare, sunlight, sunbeam, sunburn, suntan, sunstroke, and spicules.
3. *Possible answers:* nova, Mercury, Saturn, sunbeam, comet, satellite, Polaris, Aries, Taurus, Lynx, Phoenix, and galaxy.
4. Answers will vary.
5. *Possible answers:* eclipse/ellipse, solar/polar, planetoid/asteroid, Aquarius/Sagittarius, Polaris/antares, radiation/constellation, and so on.

Section 16

OCEANOGRAPHY

16-1 What Does the Sketch Represent?

Student responses will vary. Here are some possible responses for the sketches:

1. Sandy C (sandy sea).
2. Echolocation.
3. Tentacles (ten . . . tacles).
4. The loan shark.
5. Sole brothers.
6. High and low tides.
7. Rising currents.

16-2 Can You Solve the Problem?

1. Crab, barnacle, algae, fish, and anemone.
2. One - tuna fish (only one complete spelling).
3. Caribbean would then become Aribbean.
4.
5. Neal Crab.
6. S - sand, salt, shale, and slate; E - eroded particles; D - dirt, diatoms; I - iron particles; M - mineral and metal particles; E - earth particles (crustal materials), ejecta (volcanic products); N - nonmetal materials; T - till, tailings; S - silicates.
7. The lumps are nodules. Manganese, nickel, and iron materials make up the contents of nodules.
8. GU + YOT = GUYOT.
   NOD + ULES = NODULES.
   BEN + THOS = BENTHOS.
   NEK + TON = NEKTON.
   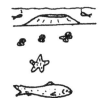
9. Dolphin, squid, whale, and turtle.
10. Sponge, coral, anemone, and oyster.

OCEANOGRAPHY (continued)

11. Dinoflagellates and radiolarians.
12. a. Ocean;
    b. Seagull.
13. Ocean fl<u>oor</u>, d<u>eep</u>, seaw<u>eed</u>, bal<u>een</u>, r<u>eef</u>, z<u>oo</u>plankton, and gr<u>een</u> turtle.
14. Aby<u>ss</u>al plain, ki<u>ll</u>er whale, transmi<u>tt</u>er (sonar), bri<u>tt</u>le star, cu<u>rr</u>ent, ve<u>ss</u>el, and fi<u>dd</u>ler crab.

16–3 What's in a Name?

1. Scales - whales; plarks - sharks; sweils - eels; snerch - perch; Slab - crab.
2. Answers will vary.
3. Shore - floor, door; trench - bench, wrench; sea - tea, key; reef - beef, leaf (table); fish - dish; whale - pail, nail; beach - peach; jetty - spaghetti; clam - jam; boat - oat, coat.
4. The Wizard of Ooze.
5. Ran, go, peg, harp, pare, pray, groan, race, pay, yap, and so forth.
6. Ann, Anne, Anna, and Ona.

16–4 Create-A-Comment

Student responses will vary. Here are examples of comments or statements:

1. "It's atoll! What else?"
2. "They are obviously forams." *Note:* This is a tough one unless students are familiar with foraminifera and know *forams* is an abbreviated from of foraminifera.

16–5 Riddle Bits

1. Centigrade scales. There are several sets of 0 to 100.
2. A sturchin, a whole, a sham, a pelt, a wallop, a squark, a wheel, and a flake.
3. Cari<u>bbean</u>.
4. Because of loss of herring.
5. C<u>oral</u>.
6. He was hoping to see some change in the tide.
7. <u>Hammer</u>head sharks and s<u>nails</u>.
8. S<u>hip</u>.
9. C<u>rest</u>.
10. Tr<u>ough</u>.

Teacher's Key

OCEANOGRAPHY (continued)

16–6 What Goes Where?

1. die (–e) = di + atom = diatom.
2. neck (nek) + ton = nekton.
3. Les + du + no = nodules.
4. fish + tu (2) + na = tuna fish.
5. The seven C's (seas).
6. C life (sea life).
7. Answers will vary.

16–7 Creative Potpourri

1. Water, water cycle, wave, weather, whale, and whelk.
2. Responses will vary.
3. Shrimp, tuna, sardines, clams, oysters, herring, crab, and so on.
4. Responses will vary.
5. Responses will vary.
6. Responses will vary.
7. Responses will vary.

LIFE SCIENCE

Section 17

SIMPLE LIFE

17–1 What Does the Sketch Represent?

Student responses will vary. Here are some possible answers:

1. A cell in a cell.
2. A pair of paramecium.
3. A square cell squared.
4. Yeast feast.
5. Old mold.
6. Simple life made difficult.
7. Silly ciliate.
8. A euglena waving the Ellum flag (or the flag ... ellum).

17–2 Can You Solve the Problem?

1. Fungi—missing letters: ng; yeast—missing letters: ye.
2. Mrs. X's last name is Euglena. The two remaining names are Glen and Lena.
3. a. coccus; b. bacillus; c. spirillum.
4.
5. Algae is plural, consisting of more than one. Take away the *e* (algae) and you'll have *alga*, a singular form or only one.

17–3 What's in a Name?

1. Answers will vary. Here are some possibilities; diatoms - Di Adams; amoeba - Amy Bah; budding - Bud Ding; Virus - Vi Russ; Moneran - Mona Ran; decomposer - Dee Compozer, rickettsia - Rick Etzia.
2. Yes, he's being truthful. His name indicates the following: slime mold - I'm old and slime - slim. Mr. Slime Mold is slim and old.
3. Lichen - A lichen is an organism made up of algae and fungi. It lives on rocks and dead trees.
4. Vorticella - a *one-celled* ciliate.

SIMPLE LIFE (continued)

5. Only one (DINOFLAGELLATE).
6. Rickettsia, radiolarian, rotifera, red algae, rust, and so on.

17–4 Create-A-Comment

Student responses will vary. Here are two possible answers:

1. "It seems like I'm always buying shoes for Si."
2. "My goodness, Edna. Darlene has grown two feet since I saw her last."

17–5 Riddle Bits

1. Lichen (lichen).
2. Both organisms have gills.
3. The sun would rise in the "yeast."
4. Diatomaceous Earth.
5. 50,000 pounds. Each plankton weighed one ton (plank*ton*).
6. A protista (Pro Tista).
7. Cellmates.
8. With cellular phones.
9. A saprophyte (*sap*rophyte).
10. To drive on.
11. A nucleus and six other organelles (organ-l's).
12. Club Fungi.

17–6 What Goes Where?

1. plasm + rp + to + o = protoplasm.
2. ates (eights) + l + flag + el = flagellates.
3. a + cell + ti + vor = vorticella.
4. ria + act + b + e = bacteria.
5. le + nu (new) + c + us = nucleus.
6. i (eye) + r + us + v = virus.
7. on + fiss (fish) + i ("aye") = fission.

SIMPLE LIFE (continued)

17–7 Creative Potpourri

1. *Possible answer:* the first sign of life.
2. Sardines - sar¢ødines.
3. a. silly + yachts = ciliates; b. diet + Tom = diatom; c. four + ham = foram; d. like + Ken = lichen.
4. a. spores; b. chitin; c. protists; d. cocci.
5. Fungicides (fung<u>icides</u>).
6. a. centriole; b. capsids; c. pellicle.
7. Responses will vary.

Section 18

PLANT LIFE

18–1 What Does the Sketch Represent?

Responses will vary. Here are seven possibilities:

1. Where botanists go to socialize.
2. Spanish moss.
3. Seed cone.
4. Tulips (two . . . lips).
5. Anther (ant . . . her).
6. Three dicots.
7. Vascular bundle.

18–2 Can You Solve the Problem?

1. Chl*oro*phyll (. . . oro . . . means gold in Spanish).
2. Bry-o-phytes.
3.
4. *ten*drils

Teacher's Key

PLANT LIFE (continued)

5. Twenty-two (twenty *tu*bers).
6. Xylem, phloem, epidermis, and meristem.
7. FOLIAGE.

18–3 What's in a Name?

1. a. vascular, palisade stomata.
   b. epidermis, pollination, turgidity, biennials.
   c. monocot, fibrous root.
   d. herbaceous stem, heterotrophic angiosperm, lower epidermis.
   e. taproot, tracheophyte, stomata, gametophyte.
2. P - petiole; L - lenticel; A - anther; N - nodes; T - taproot or tendril; S - stomata.
3. Pie, die, dip, pip, imp, dim, red, sir, sip, rip, did, mid, see, and so on.
4. Conifer - Connie Fehr; monocot - Mona Kaut; chlorophyll - Cora Phill; anther - Ann Thur; pollinate - Polly Nate; meristem - Mary Stehm.
5. Pollination - in, at, or on; conifers - on, if; internode - in, no; stomata - to, at; venation - at, on.
6. Riversnorts - liver worts; forcepale - horsetail; moonaspur - juniper; spysleds - cycads; rinkmoes - gingkoes; swaypull - maple.

18–4 Create-A-Comment

Students responses will vary. Here are two possibilities:

1. "Look, Arnold. Now there's a real dandelion."
2. "Wow! Talk about tropisms!"

18–5 Riddle Bits

1. Ra<u>dish</u> - dish.
2. <u>Bee</u>t - bee.
3. An S.O.S.: the pistil, female part of a flower, contains a <u>s</u>tigma, <u>o</u>vary, and <u>s</u>tyle, hence, an S.O.S.).
4. Sp<u>ore</u> plant - ore.
5. Liverwort.
6. They are both horsetails (horse . . . tales).
7. Club mosses (club . . . mosses).

PLANT LIFE (continued)

8. They want to be *fronds* (friends).
9. He handled only spore cases.
10. Poll*ination* (polli . . . nation).
11. To hold one or more scoops of ice cream.
12. Because they have more *bark*.

18–6 What Goes Where?

1. dril + s + ten (10) = tendrils.
2. ple (pole) + ma (May) = maple.
3. cu ("q") + le + tic = cuticle.
4. ti ("t") + pe ("p") + ole = petiole.
5. u ("oo") + it + fr = fruit.
6. plant ovary.

18–7 Creative Potpourri

1. *Vegetable 1* - potato; *vegetable 2* - carrot.
2. Eel grass, horsetail, dandelion, horse chestnut, beech, Venus flytrap, m*ono*cot (ono, a Hawaiian fish), and so on.
3. Nine: seven bean sketches *in* the diagram, one (seed), and one (s'd). The seed diagram doesn't count because the question asks for the number of seeds *in* the diagram, not the diagram itself.
4. Answers will vary.
5. Answers will vary.

Section 19

ANIMALS WITHOUT BACKBONES

19–1 What Does the Sketch Represent?

Student answers will vary. Here are some possible responses:

1. Octopi (octo . . . pie).

Teacher's Key

ANIMALS WITHOUT BACKBONES (continued)

2. Flatworm.
3. Roundworms.
4. Parasites (pair of sites).
5. Invertebrates (invert ... *ebrates*).
6. "Clam up!"
7. *Hydra*sphere (similar to hydrosphere).
8. A regenerating starfish.

19–2 Can You Solve the Problem?

1. Corals and jellyfish.
2. Tar, tear, tore, tot, to, ten, toe, tool, tea, tee, teal, tent, ton, and so on.
3. G - gnat, dragonfly; R - earwig, termite; A - mayfly, fleas; S - weevils, flies; S - bees, beetles; H - roach, moths; O - louse, locusts; P - thrips, wasp; P - punkies, scorpion flies; E - scale insects, butterfly; R - cricket, leafhopper.
4. a. 6; b. 2; c. 1. The darkened spaces reveal BONE.
5. Rhino - echinodermata.
6. Three complete spellings, one partial spelling. The final count should be four. Since starfish can regenerate lost parts, it can replace the missing *h*.)

| S | T | A | R | F | I | S | H | S | T | A | R | F | I | S | H | S | T | A | R | F | I | S | H | S | T | A | R | F | I | S |

7.
```
            e
            n
            d
            o
            s
            k
exoskeleton
            l
            e
            t
            o
            n
```

ANIMALS WITHOUT BACKBONES (continued)

19–3 What's in a Name?

1. Abe Doman, Max Illee, Kara Pace, Ross Drum, Cora Neeah, Ona Tideeum, Tim Panom, and Hal Tiers.
2. The letters *Tod* can be found in the hydra's ec<u>tod</u>erm; the letters <u>Mat</u> are located in the hydra's ne<u>mat</u>ocysts.
3. Ann, Anna, Anne, Ida, and Nelda.
4. a. spiders, caterpillars; b. spiders, caterpillars; c. worm; d. crab, crayfish, lobster; e. tick, flea; f. sponge; g. oyster, clam; h. hopping insects.
5. Cicada, ascaris, scallop, caterpillar, scale insects, caddis flies, and so on.
6. Ter<u>mite</u> and walking<u>stick</u>.
7. Bee - tree; ant - pant; flea - me; cricket - wicket; bug - mug; fly - pie.

19–4 Create-A-Comment

Answers will vary. Here are two examples:

1. "No doubt about it, Julie: that's an invertebrate!"
2. "Wow. A real spelling bee!"

19–5 Riddle Bits

1. <u>Crickets</u>.
2. When the organism is a jellyfish or crayfish or starfish.
3. Ant - <u>ant</u>enna.
4. Book louse.
5. Snails. The last five letters in snails is nails. Five out of six equals approximately 83 percent.
6. T*ape*worms.
7. F*lies*.
8. *Barn*acle.
9. *In*.
10. Unhappy sounds made by ancient Egyptian rulers.
11. Behavior (bee . . . havior)
12. A fluke.
13. Both produce ticks.

Teacher's Key

ANIMALS WITHOUT BACKBONES (continued)

14. Both of them are spongy.
15. Pill bugs and cater*pill*ars.

19–6 What Goes Where?

1. in + tri + ch + a = trichina.
2. nail + s = snail.
3. *second word:* chin + ru = urchin; *first word:* "c" = sea. *First/second* word = sea urchin.
4. ates (eights) + coel + enter = coelenterates.
5. (rust) + c + ans + ace = crustaceans.
6. su + naut + i + l = nautilus.

19–7 Creative Potpourri

1. Answers will vary. Here are some examples:
   Porifera—"I shouldn't go porifera (for if I) don't, my mom will—will be mad."
   Nematode—"Yes, that's what nematode (Emma told) us."
   Katydid—"Katydid (Katy did) all she could to help."
   Trichina—"Do a trichina." (trick, Ina)
   Mussel—"We have too many apples. We mussel (must sell) them."
2. Answers will vary.
3. a. anemone; b. sponges; c. ascaris; d. leeches; e. mollusk; f. urchins; g. beetles. All members of the "lucky seven" have seven letters in their names.
4. Crab (1–4), scallop (5–11), and snail (12–16). These three sea critters "travel in the same circle."
5. Four spiders and one arachnid. Since spiders are arachnids, the total number of spiders is *five*.

Section 20

ANIMALS WITH BACKBONES

20–1 What Does the Sketch Represent?

Answers will vary. Here are some possibilities:

1. Primates (pri . . . mates).

ANIMALS WITH BACKBONES (continued)

2. Monotreme (mono means one . . . treme).
3. A flock of C (sea) gulls.
4. Chordata (chor . . . data).
5. The results of crossing a frog with a toad.
6. Zebra (z . . . ebra).

20–2 Can You Solve the Problem?

1. Yellow perch.
2. Amph i b i a
3. *Sala*man*ders*. *Sala* is on the left and *ders* on the right. Thus man is dominated at both ends.
4. Fe__Fe__Fe__Fe__Fe__Fe.
5.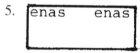
6. Bird's body ╱ head, neck, trunk, tail.
7. U. Tree has two *e*'s; true has one *e* and one *u*. Hence the difference is *u* (you).
8. M a m m a l.

20–3 What's in a Name?

1. a. robin (robbing); b. duck; c. pigeon; d. cuckoo or loon.
2. Hake - snake; snail - whale; lark - shark; seal - teal; woodchuck - duck; swan - fawn; frog - hog; mole - sole.
3. Manatee - man; elephant - ant; stickleback - tick; chickadees - chick; pigeon - pig; albatross - bat; cassowaries - ass; fowl - owl; boar - boa; a sphenodon - hen.
4. Manatee - <u>man</u>dible, <u>max</u>illa; cod - <u>rod</u>, eye structure; fowl - <u>foot</u>, b<u>owel</u>; deer - <u>derm</u>is, <u>dend</u>rite, <u>dent</u>ine; chimp - <u>chin</u>; armadillo - <u>arm</u>, m<u>axilla</u>; chipmunk - <u>chin</u>, <u>chip</u>; bear - <u>ear</u>, <u>arm</u>, <u>arte</u>ry; skunk - tr<u>unk</u>.
5. Seal, beaver, bear, anteater, pheasant, teal, eagle, beagle, seagull, Eastern brook trout, and so on.
6. Deer, goose, moose, sheep, woodchuck, woodpecker, cuckoo, loon, coon, and so forth.
7. Sentences will vary.

ANIMALS WITH BACKBONES (continued)

20–4 Create-A-Comment

Answers will vary. Here are two possible responses:

1. "Look, Mary. A game of leap man."
2. "The person who runs this place has a sick style of humor."

20–5 Riddle Bits

1. Rat (verteb<u>rate</u> . . . rat).
2. Mars (<u>mars</u> . . . upials).
3. Rodent (ro<u>dent</u> . . . dent means depression).
4. An elephant. <u>Ant</u> makes up the last three letters of its name.
5. An antelope (ant<u>elope</u> . . . elope).
6. Crocodile (cro<u>cod</u>ile . . . cod).
7. Tortoise (<u>tort</u>oise . . . tort).
8. Pouched mammals (p<u>ouch</u>ed . . . ouch).
9. A buckaroo.
10. Both have baby rattlers.
11. A skink (sk<u>ink</u> . . . ink).
12. Armadillos (ar<u>mad</u>illos . . . mad).

20–6 What Goes Where?

1. Tuatara.
2. Turkey.
3. Marmoset.
4. Parakeet.
5. Python.
6. Herring.

20–7 Creative Potpourri

1. Pun statements will vary.
2. Illustrations will vary.

ANIMALS WITH BACKBONES (continued)

3. Slogans/expressions will vary.
4. Names will vary.
5. Llama.

Section 21

REPRODUCTION AND DEVELOPMENT

21–1 What Does the Sketch Represent?

Here are some possible responses:

1. Family unit.
2. Family addition.
3. Three ovaries (three over eee's).
4. Offspring.
5. Infancy stage.
6. An undeveloped egg.
7. Life cycle.
8. A physical change.

21–2 Can You Solve the Problem?

1. Cervix, ovary, uterus, and oviduct.
2. Testes, vas deferens, epididymis, and scrotum.
3. <u>d</u>etach, bud<u>ding</u>, env<u>elope</u> = developing.
4. R - rabbit, radish; E - eel, endive; P - porpoise, petunia; R - ray, rose; O - otter, onion; D - dog, daisy; U - ungulate, urchin; C - cow, cauliflower; T - tiger, tulip; I - ibex, iris; O - ox, orange; N - nematode, nonvascular plant.
5. Em<u>b</u>ryos, umbi<u>l</u>ical, ovi<u>d</u>uct, scr<u>o</u>tum, and bl<u>o</u>od. Answer to question: blood.
6. Em<u>br</u>yo (emr), fet<u>us</u> (u), and inf<u>an</u>t (at). Unscramble *emruat* and the word is *mature*.

Teacher's Key

REPRODUCTION AND DEVELOPMENT (continued)

21-3 What's in a Name?

1. Greg Egg, Meg Egg; Mel Cell, Nel Cell; Dale Male, Gale Male; Cecil Vessel; Roy Boy; Pearl Girl, Merle Girl; Paige Age, Sage Age; Brand Gland.
2. Bert - puberty; Nancy - pregnancy; Mary - mammary; Epi - epididymis; Liza - fertilization; Liz - fertilize.
3. a. infancy; b. childhood; c. old age.
4. L - life cycle; I - infant, infancy; Z - zygote.
   A - amnion; N - nutrients, nine months; D - development;
   R - reproduction, reproduce, reproductive organs; E - embryo, eggs, epididymis; W - womb;
   S - scrotum, sperm, sex organs.
5. a. Eve; b. Ron, Ester; c. Nan, Reg, Nancy; d. Vi; e. Cal, Al, f. Rod; g. Al, Ian, Opi.

21-4 Create-A-Comment

Student responses will vary. Here are two possibilities:

1. "Look, Mark. Two zygotes!"
2. "Look at that! The poor guy's losing his coordination."

21-5 Riddle Bits

1. A balanced <u>diet</u>.
2. Inf<u>ant</u>.
3. By weighing the offspring two or three years after it's born.
4. By lying about the baby's birth weight.
5. *Five*. Mr. & Mrs. Brothers, Bill, and two sisters.
6. She delivered two Wynn's (two "wins" or twins).
7. Ten. Boyce, Sb = Boyce sounds like *boys* and Sb is the chemical symbol for tin (ten). Therefore, tin or *ten* must have been born.
8. Something about his birth *right* (to the *right* →).
9. Egg keg.
10. Mother or father.
11. Adults (adults . . . ad . . . add . . .).

REPRODUCTION AND DEVELOPMENT (continued)

21–6 What Goes Where?

1. rust + u + to = uterus.
2. LA + a + cent + p = placenta.
3. br + y + em + o = embryo.
4. s + u + feet = fetus.
5. Ne = neon or nion + am = amnion.
6. in + Ag + g = aging (first word), cess + pro = process (second word): *aging process*.

21–7 Creative Potpourri

1. Reproduce.
2. a. ovaries, cervix; b. puberty, pregnant, pregnancy; c. menstruation, testes, embryo; d. scrotum, zygote; e. epididymis, birth; f. birth, epididymis.
3. a. Utah; b. street; c. pretty; d. speed; e. moon; f. roster.
4.
    ```
              h                 s                       g
              o                 n                       o
        c o r a l         s n a i l                     r
              s                 i                 s w i n e
              e                 l                       l
                                                        l
                                                        a
    ```
5. File, I'll, Nile, rile, and so on.

Section 22

GENETICS AND CHANGE

22–1 What Does the Sketch Represent?

Here are some possible responses:

1. Cross-pollination.

GENETICS AND CHANGE (continued)

2. Ice cream clones.
3. Dominant and recessive trait.
4. A genetic cross.
5. Punnett square.
6. Gene pool.
7. Meet Gene!
8. Division of cells.

22–2 Can You Solve the Problem?

1. A gamete is a sex cell; example: egg.
2. Two sets of DNA.

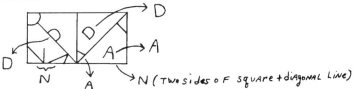

3. An autosome.
4. D__M__N__N__E__.
5. The dog: breeder. He saved *one* dollar.

22–3 What's in a Name?

1. Tame game.
2. Adenine.
3. Gregor Mendel; monk.
4. Pete Phony (phenotype); Mona Tuti (mutation); Nina Egu (guanine); Dee Anin (adenine); Lon Cing (cloning); Tom Savoe (autosome).
5. a. e - recessive; b. b - genes; c. c - pedigree; d. d - genetics; e. a - dominant.
6. hybrid - bird; clone - eon; gamete - meet; chance - cane; inherit - their; animal - mail.

22–4 Create-A-Comment

Here are two possible responses:

1. A definite "meow . . . tation."
2. "I'll bet Mendel wouldn't believe this!"

GENETICS AND CHANGE (continued)

22–5 Riddle Bits

1. Probability.
2. Gamete.
3. Because they rarely draw more than a pair.
4. Punnett square.
5. A brat.
6. Blood type.
7. The Clone Rangers.
8. Polyploidy.
9. Inbreeding.
10. A slug.
11. A french fry.
12. Stew.

22–6 What Goes Where?

1. eel + fee + f = filial.
2. edit + y + her = heredity.
3. spring + off = offspring.
4. eights (aits) + tr = traits.
5. ones + cl = clones.
6. l + mend + e = Mendel.

22–7 Creative Potpourri

1. Responses to puns will vary.
2. Answers will vary.
3. a. . . . ex*tra*, It . . . = trait; b. "BIG *ENERGY.*" = gene; c. . . . *in her, It's* . . . = inherit; d. . . . *men* de*le*gate . . . = Mendel; e. . . . *game,* "*Te*therball . . . = gamete.
4. Answers will vary.
5. *Possible answers:* genetic engineering, cloning, gene splicing, recombinant DNA, sex-link trait, protein, cytosine, thymine, guanine, adenine, offspring, inbreeding, incomplete dominance, inherit, inheritance, dominant, dominance, self-pollination, and cross-pollination.

Teacher's Key 411

Section 23

HUMAN BIOLOGY

23–1 What Does the Sketch Represent?

Here are some possible responses:

1. Kneecap.
2. Ten dons - tendons.
3. Rusty hip joint.
4. Bulging muscle.
5. Small and large intestine.
6. Nervous system.
7. Adrenalin (add-drenalin).
8. Blood vessels.

23–2 Can You Solve the Problem?

1. O, N, Se = nose; In, S, K = skin; Ne, B, O = bone; P, Li = lip; Se, No = nose; At, He = heat; Fe, Li = life.
2. The letters spell *human*.

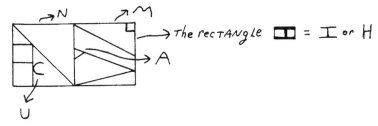

3. Use letters to spell *two red blood cells.*
4. a. Sydney; b. flood; c. root; d. hustle; e. crane; f. ingest; g. bland; h. tart; i. strain; j. flee.
5. *Possible answers:* marrow, pore, spinal cord, thyroid gland, parathyroid gland, hormones, respiratory system, circulatory system, excretory system, reproductive system, and so on.
6. a. think; b. feel; c. hear; d. see; e. smell; f. move; g. breathe; h. maintain.
7. a. pelvis; b. fibula; c. scapula; d. spine; e. tarsus; or tarsal; f. mandible; g. sacrum.

23–3 What's in a Name?

1. a. Al; b. Val; c. Lani; d. Red; e. Les; f. Di; g. Lane;
2. Descartes - Des, Art, Arte; Vesalius - Sal, Al; Galileo - Al, Lil, Leo; Crick - Rick; Mendel - Del; Redi - Red, Ed, Edi; Beaumont - Bea, Mo; Schwann - Ann.

HUMAN BIOLOGY (continued)

3. Heart, cartilage, artery, arthritis, arterial, arteriosclerosis, and so on.
4. a. spine: sublime; b. patella: umbrella; c. femur: lemur; d. tarsus: farces; e. xiphoid: typhoid.
5. Hip, eye, ear, lip, rib, toe, gum, jaw, leg, arm, cell, sac, and so forth.
6. Bone, ulna, skin, hair, anus, lung, neck, head, foot, duct, nose, pore, node, axon, cord, iris, lens, and so forth.
7. Pelvic, spine, patella, scapula, phalange, carpal, xiphoid process, and so on.

23–4 Create-A-Comment

Here are two possible responses:

1. "Now there's a strange-looking rib cage."
2. "There go two motor neurons."

23–5 Riddle Bits

1. The "funny bone" is located on the humerus.
2. Cartilage - art.
3. Both are mostly ores - pores.
4. Because they are "hip" attorneys.
5. Voluntary muscles.
6. Alimentary school.
7. A plaque.
8. The person may have a defective semilunar valve.
9. The "bronchial tree."
10. Pnemonia - "new" monia.
11. Rum - cerebrum.
12. Bones - bone.

23–6 What Goes Where?

1. east + bone + br = breastbone.
2. um + tree + a = atrium.
3. ill + aries + cap = capillaries.
4. feet + us = fetus.
5. liv + a + sa = saliva.

Teacher's Key

HUMAN BIOLOGY (continued)

6. s + ma + pla = plasma.
7. tin + a + re = retina.

23–7 Creative Potpourri

1. Responses will vary.
2. Answers: (1) asteroid, (2) comet.
   Answer: solid.
   Remaining letter is *N*.
3. Sternum, ulna, radius, carpals, tarsals, patella, ilium, sacrum, scapula, cranium, and so on.
4. enzymes - the opposite of "outzymes."; peristalsis - two "stalsis"; artery - a gathering place for artists; zygote - a funny name for an animal; intestine - the opposite of "outtestine"; ventricle - a "ricle" filled with holes; bronchus - obscene language used by rodeo horses; carpals - ride sharing buddies.
5. The ligament team won the tournament.
   The *League of Men* team won the tournament.

   Remember our slogan: "Flexor Stay Skinny."
   Remember our slogan: "*Flex or* Stay Skinny."

   I'll bile the food for the party.
   I'll *buy all* the food for the party.

   Saliva is a friend of mine.
   Sal, Iva is a friend of mine.

   Those must be nutrients.
   Those must be *new tree ants.*

   Look at the blossoms on the alimentary.
   Look at the blossoms on the *almond tree.*

Section 24

HEALTH AND ENVIRONMENT

24–1 What Does the Sketch Represent?

Here are some possible responses:

1. Barely getting enough rest.
2. Balanced foods.

HUMAN BIOLOGY (continued)

   3. Person over weight.
   4. Hormonal imbalance.
   5. Osteopath.
   6. Low on potassium.
   7. Sense receptors.
   8. Bronchial tree.

24–2 Can You Solve the Problem?

1. H - hemoglobin, hormones; E - energy, enzymes; A - adrenal glands, antibodies; L - ligaments, lymph nodes; T - thymus gland, tendron; H - heart, hypothalamus; Y - yolk (embryo), yogurt; P - potassium, phosphorus; E - endurance, exercise; R - reflexes, receptors (sensory); S - strength, saliva; O - organs, oxygen; N - neurons, nutrients.
2. *Unscrambled terms:* a. capillaries; b. arteries; c. spleen; d. oxygen; e. veins; f. heart. *Alphabetized terms:* a. arteries; b. capillaries; c. heart; d. oxygen; e. spleen; f. veins.
3. Carbohydrates - the main source of energy.
4. Calcium, iodine, sodium, phosphorus, and magnesium.
5. The two-word description is *five triangles*. An *allergen* causes an allergy - fiv<u>e</u> <u>tri</u>ang<u>les</u>.
6. Water appears *six* times; 3 *water* and 3 $\underline{H_2O}$.

24–3 What's in a Name?

1. Answers will vary.
2. a. beets; b. carrot, chard; c. beets; d. onion; e. corn.
3. a. rat, mite, deer; b. ant, nit, gnat, nag; c. bee, mice; d. eel, rat, mite, deer, cat, dog; e. ram, llama.
4. Sight, hearing, touch, taste, and smell.
5. miNeral, nUts, viTamins, exeRcise, therapIst, anTidote, vIrus, recOvery, and toxiN.

24–4 Create-A-Comment

Here are two possible responses:

1. "Now there's a REAL health nut."
2. "The disease is definitely spreading."

HEALTH AND ENVIRONMENT (continued)

24–5 Riddle Bits

1. Glucose (glue . . .).
2. Three *squares* a day.
3. Fresh flowers.
4. An Egyptian who practices the art of bone alignment. Egyptian? Sure, "Cairo" practor.
5. Bronchitis.
6. Measles (me . . .).
7. Phlebitis ("flea bite us").
8. Shingles.
9. Atrophy (a trophy).
10. Blood count.
11. A butterfly in the stomach.
12. She didn't want to be called "Ma Laria."

24–6 What Goes Where?

1. Fe + O = rust (–t) = rus + v + eye (i) = virus.
2. curvy + s = scurvy.
3. ts + tree (trie) + new (nu) + n = nutrients.
4. fish (fic) + de + in (ien) + cy = deficiency.
5. o + path + ns + ge = pathogens.
6. t + in + er + f + on + er = interferon.

24–7 Creative Potpourri

1. a. blood flood - excessive bleeding; b. sneeze disease - bad cold; c. diet riot - weight loss; d. calorie gallery - the kitchen; e. fatigue rig - exercise bike; f. immune boom - everybody's healthy.
2. Adam's apple - a piece of fruit in Adam's lunch box.

    hypnosis - a bone condition in the pelvic area.

    pharynx - an "ynx" of medium quality.

    paranoid - two unhappy people.

    keratin - a metal box used to store carrots.

    eardrum - a type of West Indies liquor made from fermenting molasses.

HEALTH AND ENVIRONMENT (continued)

3. Poison - on, no, so, is; arthritis - is, it, ha, ah; adrenaline - in, an; immunity - it, in; skeletal - at; pathology - lo, go, at.
4. Pathogen - halogen; nutrition - friction; immune - legume; disease - bees.
5. Bunion, bronchitis, boil, and bursitis.